Space Education & Strategic Applications

Also from Westphalia Press
westphaliapress.org

SE:SA
SPACE EDUCATION & STRATEGIC APPLICATIONS

VOLUME 1, NUMBER 1 • SPRING / SUMMER 2020

Melissa Layne, Ed.D. *Editor-in-Chief*

Westphalia Press
An imprint of Policy Studies Organization

Space Education & Strategic Applications
Vol. 1, No. 1 • Spring / Summer 2020

Westphalia Press
An imprint of Policy Studies Organization
1527 New Hampshire Ave., NW
Washington, D.C. 20036
info@ipsonet.org

ISBN: 978-1-63391-914-3

Cover and interior design by Jeffrey Barnes
jbarnesbook.design

Daniel Gutierrez-Sandoval, Executive Director
PSO and Westphalia Press

Updated material and comments on this edition
can be found at the Westphalia Press website:
www.westphaliapress.org

TABLE OF CONTENTS

cont'd.

EDITORS RECOMMENDED READINGS

Geopolitics of Outer Space. *Westphalia Press,* 2019

Ilada Aydin

The author examines the 21st century space technological developments and its potentials to alter nations relationships based on security considerations. This concise easy to read text brings forth the US-China relationship in this new developing field of competition.

Vindication Of Cosmic Biology: Tribute To Sir Fred Hoyle (1915-2001)

Chandra Wickramasinghe, (Editor), 2018.

In the year 2015, 100 years after Fred Hoyle was born, the ideas relating to the cosmic origins of life are slowly gaining credence in scientific circles. Once regarded as outrageous heresy, evidence from a variety of disciplines—astronomy, geology, biology—is converging to support these once heretical ideas.

This volume opens with recent review articles pointing incontrovertibly towards our cosmic heritage, followed by a collection of published articles tracing the development of the theory throughout the years. The discovery that microorganisms—bacteria and viruses—are incredibly resistant to the harshest conditions of space, along with the detection of an estimated 144 billion habitable planets around other star systems in our galaxy alone, makes it virtually impossible to maintain that life on one planet will not interact with life elsewhere. The emerging position is that life arose exceedingly rarely, possibly only once, in the history of the cosmos, but its subsequent spread was unstoppable. "Panspermiology" can no longer be described as an eccentric doctrine, but rather is the only doctrine supported by an overwhelming body of evidence. Fred Hoyle's work in this area may in the fullness of time come to be regarded as his most important scientific contribution.

Engineering, Life Sciences, and Health/Medicine Synergy in Aerospace Human Systems Integration: The Rosetta Stone Project.

Richard S. Williams and Charles R. Doarn (Editors).
NASA/SP-2017-633, HQ-E-DAA-TN51162.

A concise review of the human in the loop for space stems design, bringing together relevant examples from the past the design of aerospace systems and ensuing human factors problems.

Sean Mulligan, *Virginia Commercial Space Flight Authority*

Hank Voss, *NearSpace Launch, Inc.*

Joyce Winterton, *NASA GSFC Wallops Flight Facility*

Dale Nash, *Virginia Commercial Space Flight Authority*

Brian Crane, *Virginia Commercial Space Flight Authority*

Matt Orvis, *NearSpace Launch, Inc.*

Joseph Zeman, *Missouri State University*

Mark Peters II, *Technica Corporation USAF, Retired*

Ivan Gulmesoff, *American Public University System, [Security and Global Studies]*

Jonathan Pitt, *Virginia Polytechnic Institute and State University*

Edward Colbert, *Virginia Polytechnic Institute and State University*

Jared Clements, *Applied Technology Associates (ATA) Aerospace*

Tyler Murphy, *Applied Technology Associates (ATA) Aerospace*

Lee Jasper, *Space Dynamics Laboratory*

Charlene Jacka, *Air Force Research Laboratory/ RVEN*

Brenda Dingwall, *NASA GSFC Wallops Flights Facility*

Robert Twiggs, *Twiggs Space Lab, LLC*

Matt Craft, *Twiggs Space Lab, LLC*

SPACE EDUCATION AND STRATEGIC APPLICATIONS JOURNAL

Scope

SESA is a peer reviewed, multidisciplinary, open-access journal intended to serve and inform the international space community of the most recent advances in Space Education, Space Research, and Space Applications. *SESA* is published bi-annually, is open-access, and an appropriate venue for works-in-progress, conference papers, dissertations, and theses. Graduate research can be accommodated following peer review Book Reviews, Editorials, Commentaries and International Manuscripts are welcomed.

SESA is available in digital and printed formats.

Who Can Submit?

Anyone may submit an **original** article to be considered for publication in SESA provided he or she owns the copyright to the work being submitted or is authorized by the copyright owner or owners to submit the article. Authors are the initial owners of the copyrights to their works.

Guidelines for Submitting Manuscripts

Manuscripts should comply with the guidelines. All manuscripts submitted will be peer-reviewed by appropriate discipline experts for currency, originality, and relevancy to advancing knowledge and benefiting readers interested in the fields of space education, research, and applications. The following are some areas, but not limited to, of interest:

1. *Astronomy and astrophysics*
2. *Planetary sciences and search for life in the Solar System*
3. *Space security and defense*
4. *Satellites, with applications to weather forecasting, disaster management, navigation archeology, agriculture, earth and atmospheric science*
5. *Communications, navigation, artificial intelligence, interplanetary Internet, and health informatics*
6. *Space debris and asteroid tracking presenting threats to life on Earth and in Space*
7. *Life in the universe and habitable planets*
8. *Human space flight and emergency rescue*
9. *Space medicine, crew and space traveler health and life support systems*
10. *Human spaceflight simulations, training, crew selection and analogs*
11. *Space policy, economics, law and international conventions*
12. *Space operations, logistics and architecture*

13. *Space systems design and bioengineering [including robotics and artificial intelligence]*
14. *Planetary protection*

Original research and review of the paper should follow the APSA style guidelines [see https://connect.apsanet.org] and all biomedical papers should conform to the ICMJE "Recommendations for the Conduct, Reporting, Editing and Publication of Scholarly Work in Medical Journals" [www.icmje.org]. Recommended guidelines for the manuscript preparation should follow the following outline.

Abstract

Abstract is a summary of the paper with the same headings and 2 to 3 sentences for each section [purpose, background, etc.,].

Objective [purpose]

One or 2 sentences [start with "to explore or discuss, or review, etc.]

Background

A short introduction reviewing the current knowledge base and historical background [with references]. Describe what is known and why it is important.

Methods

Methods include: search words and key terms, search engines, manuscript inclusion criteria [language, source, year, types of publications, etc.] rating approach, i.e. subjective rating [see discussion paper]. Note that the analysis of the strength of evidence is qualitative and subjective.

Depending on the type of the submission [research, review, book, opinion, editorial etc.] methods used may be quantitative, quantitative, comparative, mixed, collaborative, action-oriented, and/or critical in substance.

Results

Cite the number, type and the strength of evidence of referenced publications. Information can be presented in a table format with the summation of the literature presented in columns: First author, Name of the publication, Method used [can substitute type of publication, such as systematic reviews, original research, etc.], Results, and Strength of Evidence. Provide info on the literature searches, which includes the number of references meeting the inclusion criteria, and their strength of evidence: the robustness will be defined by a qualifier such as: strong, good, fair, poor or not available].

Discussion

Should address and describe the salient findings of the surveyed publications and existing knowledge base, including

- *The prevailing political climate (globally and in the region/country under consideration),*
- *Ethical dilemma associated with proposed solutions,*
- *The legislative background of the topic under consideration and potential policy/political solutions.*

Conclusion

Summarize the strength of the evidence and/or applicability to practice, standards, policies or system design.

Limitations and Bias

Primarily describing the adequacy of the knowledge-base, concerns with findings and recommendations, and research gaps as appropriate.

References/Citations

Most recent peer-reviewed references on the subject matter meeting the inclusion criteria; minimize the use of non-peer reviewed citations including websites.

Conflicts of Interest

Regardless of the type of submission, all authors should provide information on potential conflicts of interest such as sources of funding, prior to submission of the manuscript to other publishers, prior presentations, etc.

Call For Papers / Publication Dates

Call for Papers *(ongoing)*

Spring/Summer (April)

Papers due: February 1

Peer reviews: February 2 – 28

Author edits: March 1 – 30

Publication: April 15

Fall/Winter 2020 (October)

Papers due: August 1

Peer reviews: August 2 – 30

Author edits: September 1 – 30

Publication: October 15

Where to Submit Your Manuscript

To submit your manuscript, please send to ***Dr. Melissa Layne*** mlayne@apus.edu

Editorial

Melissa Layne
American Public University System

The editors and advisors, together with the American Public University System(APUS) and in collaboration with the Policy Studies Organization (PSO), are pleased to introduce the first edition of the biennial Space Education and Strategic Applications (SESA) Journal. Our goal is to inform the industrial, military, education, and civilian sectors of advances in Space Education, Space Research, and Space Applications. More than ever before, our world is developing a focus upon and is enthusiastically supportive of these advances. This excitement was part of our new, "fourth" industrial revolution—one which provided us a glimpse of what was yet to come in the space industry, as well as a rekindling sense of community.

Unfortunately, an ominous cloud blanketed the earth, abruptly silencing this enthusiasm in November 2019. By many accounts, this event was anticipated by scientists and health organizations for some time, however no one was prepared for this. Our newfound focus on space was called back to earth and forced to re-examine biology. More devastating, our sense of community was immediately extinguished. The human mind and body suddenly became hostage to a force that many of us have never before experienced. This is not a natural disaster—we have been immobilized by an unforgiving COVID-19 pandemic. Kate Brown, MIT professor and author of *The New Yorker* article, "The Pandemic is Not a Natural Disaster", poses some important questions to consider,

> "In the midst of the Coronavirus outbreak, this idea of a body as an assembly of species—a community—seems newly relevant and unsettling. How are we supposed to protect ourselves, if we are so porous? Are pandemics inevitable, when living things are bound so tightly together in a dense, planetary sphere?"

The rapid, global spread of the COVID-19 pandemic have infected millions of people and have cost hundreds of thousands of lives. We've endured several months of quarantine during which faith, trust, hope, and community has also been lost. Across the globe, our physical and mental well-being has been put to the most challenging of tests. The term "unprecedented" will forever describe this period in history.

However, from this catastrophic human crisis, our generations have witnessed faint glimmers of light emerging; representing the beginnings of our healing process. While sequestered in our homes, we were forced to slow down and appreciate life, our families, and our friends. We became extremely creative by continuing to communicate in new ways, run our businesses by using delivery and pick-up services, holding online classes and virtual graduations for students, and allowing employees to work from home. We have not surrendered to this virus and will not allow it to squelch the human spirit nor our newfound creative spark.

As if a pandemic wasn't enough, on May 25th, the unjustified murder of George Floyd, a 46-year-old black man, died with his neck under the knee of a white law enforcement officer in a horrific, caught-on-tape incident. The world watched the tape over and over again in disbelief rhetorically and emotionally asking why? This is a pandemic of a different kind. We learned the depth with which racism, inequality, and discrimination is embedded in our everyday lives. Whether "asymptomatic" or "symptomatic" this disease is long overdue for the development of a vaccine and was poised to surge in numbers. The response to this particular pandemic has been different to the COVID-19 response. Significant changes around laws are occurring across the country such as removing clauses that protect officers with prior disciplinary records, changes in HR training, curricula modifications in LE academic programs. Most importantly, it's prompted something long overdue–deep and perhaps uncomfortable conversations between blacks and whites. We are definitely a different nation since May 25th.

Five days later, on May 30th, 3:22 pm, our emotional roller coaster took another swift turn–finally, in a positive way. Although we were encouraged not to attend physically, another event renewed our excitement and sense of community. Space made history by returning to the launchpad with NASA and SpaceX's spaceship Crew Dragon, the first commercial spacecraft to launch American astronauts, Doug Hurley and Bob Behnken into Earth's orbit. This event represented the indestructible immunity and "superhuman" resilience to global disasters of any kind. Rather than wasting narrative describing the event, I encourage you to view the images capturing the incredible story leading up to this historic launch in the article, "Images Capturing the Historic SpaceX Demo-2 Launch." As prefaced in the article, "a picture is worth a thousandwords."

The military, education, industry, and civilian sectors are partnering to further space initiatives. We also hope the incredible work our authors share with us to share with you, also keeps our sense of community growing and thriving.

Melissa Layne, Ed.D.

Editor-in-Chief, SESA

Introduction to Our Featured Article

Melissa Layne
Editor-in-Chief, *Space Education and Strategic Applications*

Weeks before the news of COVID-19 surfaced, the author of our featured article and also whom this special inaugural issue of *Space Education and Space Applications* is dedicated, published the following correspondence entitled, "Space Weather and pandemic warnings?":

> *"On the basis of this data, there appears to be a prima facie case for expecting new viral strains to emerge over the coming months and so it would be prudent for Public Health Authorities the world over to be vigilant and prepared for any necessary action. We need hardly to be reminded that the spectre of the 1918 devastating influenza pandemic stares us in the face from across a century." Current Science, on November 25, 2019:*

From what data did this scientist base this ominous, forthcoming pandemic? How could anyone possibly foresee something of this magnitude coming? Who is the person behind such a bold prediction?

Before hastily revealing the identity of the internationally renowned Sri Lankan-born poly-math scientist who made this prediction, and who, in his 81st year has stood strong in defense of his scientific discoveries, I would like to bring to the fore some general attributes among scientists, the scientific community, and the public at-large. This elaborate introduction tailored for our author is well-deserved, necessary, and timely to the publication of this issue; for we are not simply publishing an inaugural issue, we are documenting a time in which incredible things are happening *around* us and *to* us. Events that are absent of clear explanations, are contradictory, or that are explained based upon ill-defined logic. Our author proposes explanations that have been flatly rejected and ridiculed in the past, but are increasingly being supported by growing evidence. **As a teaser, he provides us with logical explanations around COVID-19 and other past pandemics. In fact, in an article published this past November 25, 2019, he actually *predicted* the Coronavirus pandemic.**

Therefore, juxtaposed against this chaos we have the rare opportunity to witness the transformation of a long-held scientific paradigm, to a paradigm which starkly contradicts it. This is a rare moment in history comparable to scientist Arnold Sommerfeld's assertion that the electron orbits elliptical and adjusts to Einstein's theory of relativity. Although we know now that electrons *do not* orbit the nucleus at all, early 20th century scientists were convinced this was true for many years. However, later evidence debunked his theories. I truly believe that we will witness a paradigmatic change on how we view viruses, bacteria, and pandemics, thanks to the remarkable work of our featured author.

Now, a little bit about what makes scientists tick.

The Curious Mind

The *thirst* for knowledge stems from one source ... curiosity. The human race is prewired with curious instincts at birth. Behind every discovery—no matter how simple or complex, there exists a curious mind. Scientists, in particular, usually aim their curiosity toward finding solutions, making connections, seeking understanding with things that cannot be explained, or advancing already-established theories.

Research scientists such as Galileo, Aristarchus, Copernicus, Sagan, Einstein and others have generously contributed to the scientific community with discoveries that have changed our lives, life around us, and life forthcoming. Their work has provided a foundation from which subsequent scientists have expanded and advanced. They are models of lifelong curiosity, careful observation, relevant questioning, fierce exploration, and fervent searchers of truth.

The Really Valuable Factor is Intuition ~ *Albert Einstein*

The particular scientists mentioned above were also considered what the science community calls "speculative" or "intuitive" scientists - those who are inspired by an experience, or who may not be necessarily thinking about their respective topic, but that the reasoning or solution to their questions suddenly appear, as if out of nowhere during certain circumstances. For example, one of the greatest mathematicians, Srinivasa Ramanujan, is a model of curiosity, intuition, and unwaning determination. This self-taught genius had an immense fixation that followed him until his death: the number pi (π). As Ramanujan fell asleep, he had recurring dreams about Namagiri, a Hindu goddess.

Interestingly, Ramanujan claimed that the Hindu goddess would present formulas, equations, and theorems to him in his dreams every night. When he would wake up the next morning, he would write down what he could remember from the dream:

> *"While asleep, I had an unusual experience. There was a red screen formed by flowing blood, as it were. I was observing it. Suddenly, a hand began to write on the screen. I became all attention. That hand hand wrote a number of elliptic integrals. They stuck to my mind. As soon as I woke up, I committed them to writing." ~ Srivivasa Ramanujan*

After a few years, he had accumulated 3,900 formulae. However, because he was unable to show how he arrived at these equations and formulas (as they need to be replicable for other scientists to follow) Ramanujan's "dream" methodology caused him much opposition from the scientific community.

The origin of Albert Einstein's "hunch", later known as his *Theory of Relativity*, allegedly came to him while sick in bed.

Astronomer, astrophysicist, astrobiologist, and best known for his research on extraterrestrial life, Carl Sagan, recounts an experience as a child that confirmed his life as a scientist. Upon convincing his mother to get him a library card, he shares what he did next:

> *"I went to the librarian and asked for a book about stars; ... And the answer was stunning. It was that the Sun was a star but really close. The stars were suns, but so far away they were just little points of light ... The scale of the universe suddenly opened up to me. It was a kind of religious experience. There was a magnificence to it, a grandeur, a scale which has never left me. Never ever left me." ~ Carl Sagan*

Similar to the researchers before him, Sagan was "an 'ideas person' and a master of intuitive physical arguments." (Morrison, 2007).

A Life of Scrutiny and Ridicule

Although we have widely acknowledged the aforementioned scientists' exceptional discoveries as "truth", such as that earth and other planets orbit the sun, and space and time are interwoven into a single continuum known as space-time, their search for truth was often wrought with ridicule and skepticism. This form of oppositional response is so prevalent within the scientific community, it actually carries a name ~ *"attack-escape."*

Attack-escape is when *we* (scientists and nonscientists) are introduced to a novel idea that doesn't quite fit within our current belief system, we tend to judge the new discovery before bothering to look at the evidence that has been provided in support of the discovery. Innovative and revolutionary discoveries, in particular, are largely the hardest hit as they are presented against long- and firmly-established convictions. Unless the scientist continues to pursue his / her truth with determination, they will either be ignored, or sadly, not live long-enough to experience the eventual acceptance of what sparked the journey stemming from that initial curiosity. Physicist Sir Alan Walshe is spot-on with this analogy:

> *"The itch to suffocate the infant idea burns in all of us"*
> *~ Sir Alan Walshe*

If a scientist does not have the determination and courage to face his opposers and adversaries and to keep searching for his truth, this will ultimately have an emotional and mental toll that cripples and squelches any spark left in his quest.

Mentors and Colleagues

To prevent this from happening, many scientists believe it beneficial to surround themselves with a mentor or colleagues for support and motivation. For example, upon his announcement of his discovery, the X-ray, Wilhelm Rontgen faced criticism and outright abuse over the innovation. However, it was his mentor and famous physicist, J.J. Thomson, who did not falter in his conviction that Wilhelm's X-ray was revolutionary and would change the world of medical science.

Our featured author has very similar characteristics, experiences, and paths similar to the great above-mentioned minds in science.

On Curiosity

As a child, he always felt a "mysterious connection" with the universe which sparked his curiosity and interest in space. At the age of 15, he wrote the following poem:

Amongst the myriad stars
I stand alone
and wonder how much life
and love there was tonight

~ Chandra Wickramasinghe

On Intuition

"I grew up in a Buddhist culture in which our connection with the external cosmos was deeply impressed. So when I began studying science, particularly biology, which is Earth-centred in a very fundamental way I was shocked into thinking about these things - perhaps thinking in a different way to the way Western science had developed over hundreds or thousands of years."

~ Chandra Wickramasinghe

On Scrutiny

Despite thirty years of working on his and his mentor's (Sir Fred Hoyle's) theory of "Cometary Panspermia" they experienced significant scrutiny and disagreement in the astronomical, biological, and medical fields. Their evidence for "Cometary Panspermia" and disease from space was mocked, their ideas actively suppressed, and their peers abandoned them without responsibly reviewing their work. Fortunately for us and our planet, our author and his mentor did not give in to consensus, and published over 300 papers in major scientific journals, over 75 in the journal *Nature* on Panspermia and disease from space, as well as over 30 popular books.

He provides sage advice to scientists—young and older, taken from the last words of Buddha to his main disciple, Ananda:

"Be lamps unto yourselves,
Hold fast to Truth as a lamp;
Hold fast to the truth as a refuge.
Look not for a refuge in anyone beside yourselves."
~ Buddha

Without further ado, it is my pleasure to introduce the esteemed author of our featured article and developer of astrobiology.

Our Featured Author: Professor Chandra Wickramasinghe

Our featured author for this special inaugural issue of the *Space Education and Strategic Applications* journal, Professor Wickramasinghe, was born in Sri Lanka and was educated at Royal College, Colombo and later at the University of Ceylon. In 1960 he obtained a First Class Honours degree in Mathematics and won a Commonwealth scholarship to proceed to Trinity College Cambridge.

He commenced work in Cambridge on his PhD degree under the supervision of the late Sir Fred Hoyle, the iconic astronomer of the 20th century, and published his first scientific paper in 1961. He was awarded a PhD degree in Mathematics in 1963 and was elected a Fellow of Jesus College Cambridge in the same year.

In the following year he was appointed a Staff Member of the Institute of Astronomy at the University of Cambridge. Here he began his pioneering work on the nature of Interstellar Dust, publishing many papers in this field that led to important paradigm shifts in astronomy. He published the first definitive book on *Interstellar Grains* in 1967. In 1973 he was awarded Cambridge University's highest doctorate for Science, the ScD.

Chandra Wickramasinghe is acknowledged as a leading expert on interstellar material and astrobiology. In fact he and Fred Hoyle invented the word Astrobiology for the burgeoning discipline that married astronomy with biology. Chandra, over a lifetime, has made very many important contributions in these fields. In 1974, he first proposed the theory that dust in interstellar space and in comets was largely organic, a theory that was shortly afterwards vindicated and effectively led to the birth of the theory of cometary panspermia.

Jointly with the late Sir Fred Hoyle he was awarded the International Dag Hammarskjold Gold Medal for Science in 1986. Chandra Wickramasinghe was a UNDP Consultant and Advisor to the President of Sri Lanka in 1982-84, and played a key role in the setting up of the Institute of Fundamental Studies in Sri Lanka. In 1983/84 he was appointed the founder Director of the Institute of Fundamental Studies by President J.R. Jayawardene. In 1992 he was decorated by the President of Sri Lanka with the titular honour of *Vidya Jyothi*.

In 1973, he was appointed Professor and Head of the Department of Applied Mathematics and Mathematical Physics at University College, Cardiff, being the youngest Professor appointed at the University upto that time. He was responsible for starting an Astrophysics research group in Cardiff under the auspices of a new Department that was formed under his headship, the Department of Applied Mathematics and Astronomy. He remained Head of this Department until 1989 by which time the Astronomy Research School in Cardiff was regarded as being one of the best in the UK. From 1989-1999 he held the post of Professor of Applied Mathematics and Astronomy within a newly structured School of Mathematics at Cardiff University of Wales.

In the year 2000, he was appointed Director of the newly formed Cardiff Centre for Astrobiology. In 2006 he retired from Cardiff University and has since been a "Professor at Large" in a number of Universities and Institutions worldwide. He is currently Director of the Buckingham Centre for Astrobiology, University of Buckingham, and an Honorary Professor there as well. He is also Honorary Professor at the University of Ruhuna, Sri Lanka, and an Honorary Professor at the Sir John Kotelawala Defence University of Sri Lanka as well as an Adjunct Professor at the National Institute of Fundamental Studies in Sri Lanka.

He is an award-winning poet and the author and / or co-author of over 40 books and over 300 scientific papers, over 50 being in the journal *Nature*. He has held visiting professorial appointments in a large number of Universities world-wide. In recognition of his extensive contributions to science and culture he was awarded an honorary doctorate by the Soka University of Tokyo, Japan in 1996. He was awarded the degree of Doctor of Science (Honoris Causa) by the University of Ruhuna, Sri Lanka in 2004.

In 2018, he and 32 of his colleagues published the pivotal "Cause of Cambrian Explosion - Terrestrial or Cosmic?" by the highly-respected journal, *Progress in Biophysics and Molecular Biology* (not too shabby).

What is Professor Wickramasinghe Searching For?

Chandra Wickramasinghe, his mentor Sir Fred Hoyle, and many colleagues have focused on one, very important scientific conclusion:

> *"—that life was seeded here on Earth by life-bearing comets as soon as conditions on Earth allowed it to flourish (at or just before 4.1 Billion years ago); and living organisms such as space-resistant and space-hardy bacteria, viruses, more complex eukaryotic cells and organisms (e.g. Tardigrades), perhaps even fertilised ova and plant seeds, may have been continuously delivered ever since to Earth helping to drive further the progress of terrestrial biological evolution. This process, since the time of Lord Kelvin (1871) and Svante Arrhenius (1908) has the scientific name "Panspermia".*

Why is this Relevant Now?

Interestingly, we have before us two very important events occurring at the moment—a worldwide pandemic, and the other, plausible, evidence-supported explanations on the origin of this virus.

Although the COVID-19 pandemic has been described as "unprecedented" a million times over, it is not. There have been other "unprecedented" pandemics before it.

What *is* "unprecedented" is that a) we have never been able to fully explain why/how pandemics occur, or from where they come (and no, they do not come from animals, according to Chandra and Fred Hoyle); yet b) we now have in front of us, an explanation based upon evidence that makes sense, and that provides direction for further research, and the development of academic multidisciplinary programs to secure and nurture advancement in this field of study.

Thank you Professor Chandra Wickramasinghe, for sharing your journey as a scientist, your outstanding contributions to space science, and for providing us with words of wisdom that we can take with us and apply to our own respective fields. Ω

DIVINE MANIAC

There he is—our grey-haired sage
Gazing at the star-stud sky;
Pacing up and down.
Suddenly he stops to scratch his head;
Something puzzles him.
He bites his pencil in nervous agitation
And mutters a stifled curse.

At last he nods his head
And smiles to himself triumphantly—
A sudden flash of inspiration
Has perhaps enlightened him.

He pulls out a scrap of paper
And scribbles something upon it.
And then, his eyes still fixed upon the starry sky,
He continues to stand there, motionless,
As if awe-struck by its beauty.

This man is a queer sort of chap.
He can foretell eclipses.
And like a book he knows the world—
The world of stars and atoms,
And of life and love as well.

Is he a human being, or is he not?
I wonder.

Some people say he's mad;
Some think he's just a bit eccentric.
As for me, I don't know what to think;
But of this much I'm convinced—

That if he's really mad,
His madness is divine.

~ *Chandra. Wickramsinghe, 1956*

A Transformative Paradigm of Cosmic Life

Professor Chandra Wickramasinghe

> "It is far better to grasp the universe as it really is than to persist in delusion, however satisfying and reassuring ..." Carl Sagan

When I first arrived in Cambridge in October 1960 to begin working with Sir Fred Hoyle, who was to become my mentor and latter colleague and collaborator for close on half a century, he had already made a monumental discovery. Working with Margaret and Geoffrey Burbidge and William A. Fowler, Fred Hoyle had already begun to change our perception of ourselves in a most fundamental way. He had demonstrated to the conviction of the world that we are all made of star-stuff. The atoms of carbon, oxygen, nitrogen, phosphorous of which we are made were synthesised in nuclear reactions that took place in the deep interiors of stars, and expelled into space by the explosion of supernovae. Likewise, the atoms of silicon and iron in the rocks of the Earth were produced in stars. This was the first step in understanding our genesis that came to be accepted as a fact without much dissent. In this article, I shall discuss the next steps in our journey towards unravelling the nature of our existence – steps that were taken by Fred Hoyle, myself and our many students and collaborators over nearly half a century. Because we now appeared to be trespassing on many disciplines including biology that were not considered to be rightfully ours, and attempting to overturn a long-established philosophical paradigm, this part of our journey turned out to be long and tedious. Nevertheless, we continued to persevere and at last it is beginning to look as though we are on the threshold of a major shift of scientific paradigm that could be far-reaching and profound. Not only the atoms of our planet but life itself did not start on Earth but originated in the context of the vast universe of which our local cosmic niche is but a minute, infinitesimal part.

Life on Earth

The Earth teems with life of all kinds ranging from the simplest microorganism to the most complex of life forms—plants, animals, humans. We humans—*Homo Sapiens* —are now perched atop this magnificent pyramid of terrestrial life declaring ourselves to be in command of all we survey. Over the past century biologists have unravelled the mind-blowing complexity of life at the molecular level as well as its super-astronomical information content as is manifest for instance in the arrangement of amino acids in crucial enzymes. At the same time astronomers have unravelled a universe that is every bit as complex, informationally rich and as magnificent as life itself. For too long however, we have failed to appreciate that there must exist an intimate and inextricable connection between life on the one hand and the vast external universe on the other. Only by acknowledging this link would it ever be possible to fully comprehend the world in which we live.

For well over a century the idea of life starting by a process of "spontaneous generation" on the Earth in a primordial soup of organics has been firmly locked into the cannon of science. Attempts to synthesize life from non-life have continued in the most advanced biotechnology laboratories for well over half a century. With the passage of time all such hopes have turned out to be illusory. Every attempt that has been made to replicate the process of spontaneous generation in the laboratory under the widest possible range of conditions has ended in dismal failure. Thus, Louis Pasteur's 1863 dictum of over a century and a half *Omne vivum ex vivo*, (*all life [is] from life*) rings truer today than it has ever done.

Four decades ago the late Sir Fred Hoyle and I had already accumulated enough supportive evidence to eventually assert with confidence that terrestrial life must be inextricably linked to the cosmos at large. The main connecting link was comets and cometary debris that continually gains entry to the Earth's environment. Supportive evidence came from many different scientific disciplines, with the result that a majority of scientists working within confines of their own special discipline remained loathe to transgress the limitations of their particular boundaries, and so to contemplate accepting a wider cosmic world view. This position is at last beginning to change.

The most powerful single argument for life being a cosmic rather than a purely terrestrial phenomenon was articulated by the late Sir Fred Hoyle way back in 1980, summarizing the position that we had reached at the time:

"The very small probabilities, which one calculates for the assembly of these substances (e.g. enzymes), demonstrates as near to certainty as one would wish that life did not originate here on the Earth. Indeed, the infinitesimal probabilities demonstrate that life is even too complex for its origin to be confined within our galaxy alone. The resources of the whole universe were almost certainly needed ..."

If there was a deep principle of nature that drove inorganic systems towards the emergence of primitive life—the evidence for this would have long since been discovered in the laboratory, which as we noted, has not. Moreover, with calculations showing grotesquely low *a priori* probabilities for the transition from non-life to life only two options remain:

(1) The origin of life was an extremely improbable event that must have occurred on Earth against all odds (because we are here!) but will consequently not be reproduced elsewhere. In that case we would indeed be hopelessly alone as a life system in the Universe.

(2) Alternatively, a very much vaster cosmic system than was available on Earth, and a very much longer timescale was involved in an initial origination event, after which life was transferred to Earth and elsewhere by processes that the late Sir Fred Hoyle and I proposed many years ago—*cometary panspermia.*

Figure 1: *Left.* Sir Fred Hoyle and Professor Chandra Wickramasinghe, in Sri Lanka, 1982. *Right.* Professor Chandra Wickramasinghe and Sir Fred Hoyle bronze sculpture in Cambridge, 2019. *Photo Credit: Priya Wickramasinghe*

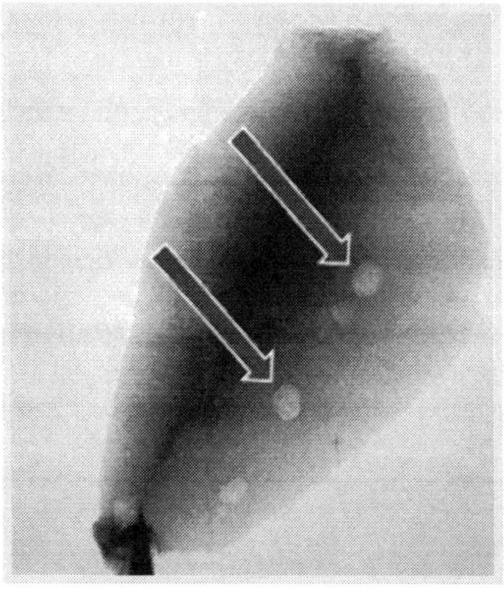

Figure 2: Carbon inclusions in a 4.1 Gya zircon crystal *Photo Credit: Stanford/ UCLA*

Predictions from the theory of Cosmic Life

From the 1980's onwards with the emergence of new areas of science and the development of new technologies, the many predictions that followed from this single unifying hypothesis of cometary panspermia came along to be tested. Remarkably, they have all survived the well-known Poperian test of consistency with unfailing accuracy. These include the discovery of extremophiles—microorganisms on Earth that are able to survive extreme conditions including the harsh cosmic environment; and space experiments that have actually demonstrated the survival of viruses and bacteria on the surfaces of rockets that were shot at high speed through the atmosphere, as well as on the surfaces of orbiting spacecraft.

Finally, we recently have the discovery that the *very first evidence of microbial life* on Earth locked away within crystals of zirconium in rocks that formed 4.1-4.2 billion years ago—and are now exposed in the *Jack Hills* outcrop in Western Australia (Figure 2). This latter discovery in my view puts paid to the possibility of any primordial soup brewing on Earth because these fossils were deposited at a time when the planet was being relentlessly bombarded by comets and meteorites.

Direct evidence from interstellar dust and comets

The first discoveries of organic molecules (even biochemicals) existing in the space between stars go back to the early 1970's when I began to work at after my PhD as a Research Fellow at Jesus College Cambridge. Since this time, the range and complexity of organic molecules in space have grown almost without limit. Sir Fred Hoyle and I were the first to point to the unmistakable biological provenance of some of these astronomically discovered molecules; but we tended to be ignored or even ridiculed for a long time. If we were right, too many paradigms and beliefs had to be shattered.

The first astronomical discovery showing interstellar (cosmic) dust to mimic the infrared absorption pattern (spectrum) of bacteria was made at our behest by Dayal Wickramasinghe (my brother who was Professor at ANU) and David Allen who used the Anglo-Australian telescope to obtain the data shown as the points in Figure 3.

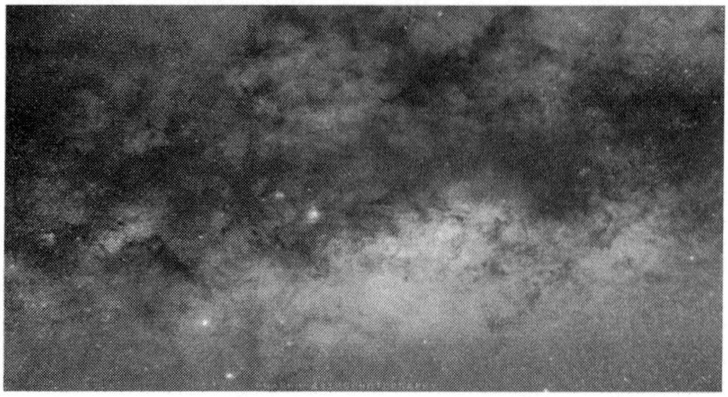

Figure 3: Dust clouds in the direction of the Galactic Centre.

This correspondence between astronomical data and prediction is reproduced below (Figure 4) and has stood the test of time to obtain. Still many scientists would prefer to suggest that this was a mere coincidence—a purely coincidental assemblage of simple organic structures that happened to mimic biology.

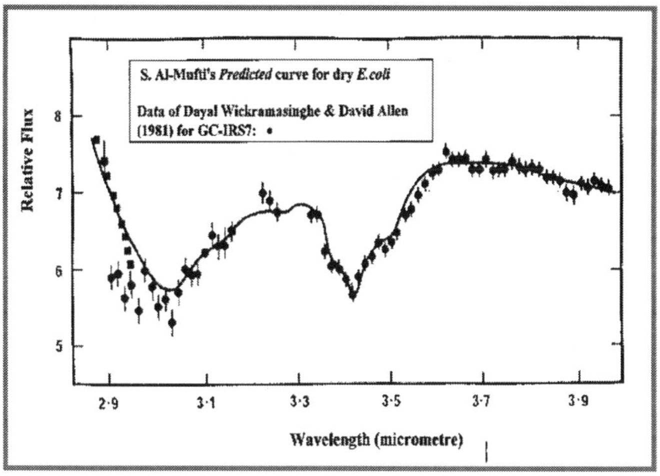

Figure 4: Points astronomical data, curve prediction of bacterial dust

Decisive evidence for complex aromatic and aliphatic carbon-based molecules (ring molecules and long chain molecules) now exists everywhere in our Milky Way galaxy, and even beyond in galaxies as far away as 8 billion light years. Whilst all such data still tends to be interpreted conservatively avoiding "biology" with the suggestion that we may be witnessing exceedingly improbable "primordial soup-type events" on a cosmic scale, an all-encompassing cosmic biology remains by far the most plausible and logical option. This is evidence of panspermia in action—the smaller organic molecules in interstellar space being degradation products of iterant bacteria and viruses.

Comets in our solar system have been the focus of several space missions since 1986 following ESA's Giotto successful mission to Halley's comet. The Giotto mission showed clearly that the prevailing theory that comets are dirty snowballs had to be abandoned in favour of comets rich in organic molecules, and most likely also containing viable bacteria and viruses.

More recent explorations of comets, culminating in the Rosetta Mission to Comet 67P/C-G (Figure 5), have yielded a formidable body of evidence, all showing consistency with the existence in comets of microbial material. Many species of fermenting bacteria are known to be able to produce ethanol from sugars, so the recent discovery that Comet Lovejoy emits ethyl alcohol amounting to 500 bottles of wine per second would appear to be a clear indication that such a microbial process is operating.

Figure 5: September 10, 2014 imaging shows jets of cometary activity along the whole neck of the comet. *Photo Credit: ESA/Rosetta/MPS for OSIRIS Team MPS/UPD/LAM/IAA/SSO/ INTA/UPM/DASP/IDA.*

The essence of the emergent theory of cosmic life is that the entire galaxy—perhaps the entire Universe—is one single connected biosphere. Physical transfer involving exchange of meteorites, comets, and dust with a consequent intermingling of genes on a cosmological scale is far more probable than independent life origination events that are still widely *assumed* without *any* proof in conventional science.

Habitable exoplanets

Figure 6: Artist's impression of a habitable planet orbiting a red giant star

In 2009 NASA launched its orbiting Kepler telescope, which was specifically designed to discover planets that are the size of Earth. The detection process involved tracking down minute blinks (dimming) in the star's light when a planet transited periodically in front of it during its orbit. At present, nearly 4000 definite as well as probable detections of habitable planets have been made within only a very small sampling volume of our Milky Way. Most of these planets orbit red dwarf stars that are on the average twice the age of our sun. Extrapolating from the sample of present detections, the estimated total number of habitable planets in our Milky Way galaxy is reckoned to be in excess of 100 billion.

These billions of exoplanets would of course be in different stages in regard to the development of indigenous and adapted life systems, and in a fraction of such planets life may even have become altogether extinct. But with the many astrophysical processes that could operate in transferring life-bearing material across galactic distances it would now seem inevitable that such habitable planets in the galaxy would be biologically interlinked into a single galactic biosphere. The processes of horizontal gene transfer that are well recognised within the context of terrestrial biology would have its widest and most natural range across the entire galaxy, and the universe at large.

The Octopus

If a single discovery is to serve as a watershed in the journey to accepting our cosmic origins, it is a recent study of two related species, the squid and the octopus. The squid has an antiquity in the geological record that goes back to the great metazoan explosion of multi-celled life 540 million years ago. The octopus apparently branches out from the squid line about 400 million years ago, presumed to evolve from an ancestral squid. Recent DNA sequencing of the squid and octopus genomes has exploded a bombshell. The squid contains a very meagre compliment of genes adequate to serve its modest survival needs. The emergent octopus, on the other hand, has ***over 40,000 genes*** (the human has only 25,000 genes) and many of these genes code for complex brain function. Others code for a highly sophisticated camouflage capability including rapid switches of colour. The octopus is incredibly more complex in structure and performance than its squid predecessor. Where did the suite of genes coding for complex brain function come from? They were not present in the ancestral squid or in any other living form that existed on the Earth at the time. The clear implication is that they came from outside the Earth—external to terrestrial biology—part of the cosmic milieu of life.

Comets, Meteorites, Micrometeorites

In our theory of Cosmic Panspermia, for which there is growing evidence, comets play a crucial role, serving as storage systems, amplifiers, and distributors of cosmic life in the form of bacteria and viruses. When a life-bearing comet makes its repeated orbits around the sun its volatile substances are pro-

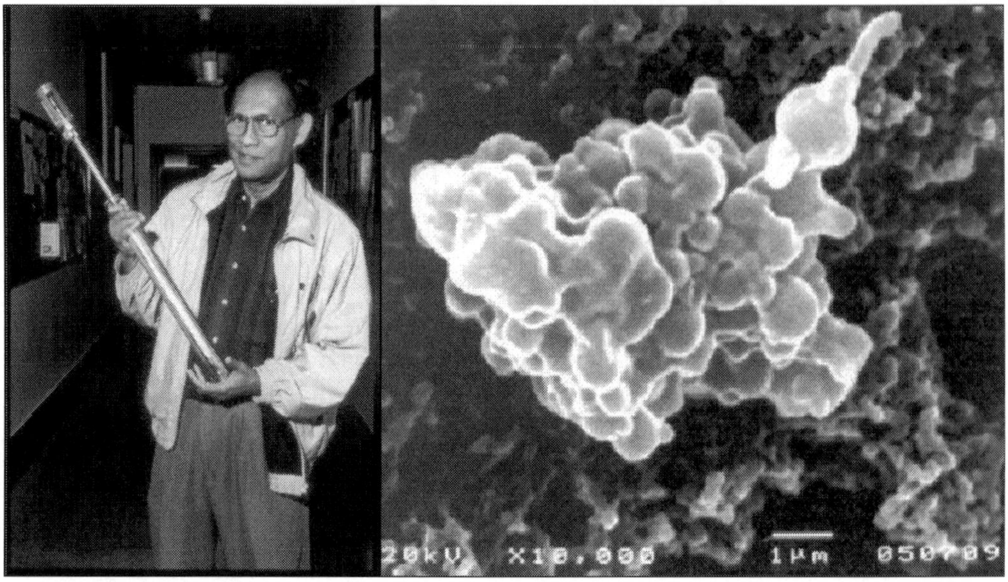

Figure 7: *Left.* Professor Wickramasinghe holding a cryogenically cooled sterilised stainless-steel cylinders. *Right.* Air being sucked into the cylinders (41km above the Earth's surface) before they were parachuted back to the Earth.

gressively vaporised and eventually we end up with what could be recognised as large carbonaceous meteorites. The number of close perihelion passages that a comet can survive before becoming completely stripped of volatiles is probably a few hundred. Carbonaceous chondrites represent fragments of comets denuded of volatiles but retaining a residue of silicates and more refractory organic structures, possibly fossilised as well as viable microorganisms. From time to time, such objects find ingress into the Earth and can be examined in the laboratory.

Infalling cometary debris

One crucial test of the theory of cosmic life is to probe the stratosphere for in-falling alien genetic systems—bacteria and viruses. To urge international space authorities with the capability of doing this was far from easy. The first dedicated effort to test the idea of bacterial in-fall from comets was carried out in collaboration with scientists at ISRO (Indian Space Research Organisation) in 2001.

Cryogenically cooled sterilised stainless-steel cylinders each with a volume of 0.35 litre were launched as part of a balloon payload. At a height of 41km the seals were opened and the ambient air sucked into the cylinders before they were parachuted back to the Earth (See Figure 7).

The seals were open in a clean room and the contents of the cylinders passed through a series of membrane filters. Positive detections of in-falling microbiota were made, and the number of bacterial cells collected in a measured volume of the stratosphere at 41km led to an estimate of an in-fall rate over the whole Earth of 0.3-3 tonnes of microbes per day. This converts to some ***20-200 million bacteria per square metre arriving from space every single day.***

This truly vast number unfortunately pales into insignificance when compared to bacteria and viruses originating from the Earth's surface, some of which could be lofted to heights of about 3km in upward air currents and brought down in mist and rain. The total flux of bacteria and viruses reaching near the Sierra Nevada mountains was recently measured by Reche et al (2018). The average flux of mainly recycled viruses was found to be ***800 million per square metre per day.***

If both the space incident microorganisms and terrestrial microbes originate from disconnected pieces of a single cosmic biosphere, their genetic difference may well turn out to be subtle and even difficult to detect. Indeed, an ISRO sponsored balloon flight into the stratosphere in 2006 recovered three new bacterial species that are genetically similar (80% homologous) to known terrestrial species but sufficiently different to be classified as a different species (Shivaji et al, 2007). The first of the new species recovered from 41km was named *Janibacter hoylei,* after Sir Fred Hoyle, the second as *Bacillus isronensis* recognising the contribution of ISRO in the balloon experiments, and the third as *Bacillus aryabhata* in honour of India's celebrated ancient astronomer.

More expensive and sophisticated investigations need to be carried out even on the samples collected so far, if we are to prove beyond doubt that these microbes are unequivocally alien.

Pandemics in history

If, as we have pointed out, a cometary impact or impacts led to the commencement of life on the Earth 4.2 billion years ago, it is reasonable to assume that subsequent arrivals of cometary material would carry biological material that would affect the terrestrial biology. Such invasions could take the form of new viral and bacterial infections that strike our planet at irregular intervals, pathogens sometimes drifting down onto the surface in the form of clumps of meteoritic material.

Reports of the sudden spread of plagues and pestilences punctuate human history throughout the millennia. The various epidemics, scattered through history and throughout the world often bear little or no resemblance one to another. However, they share the common feature of afflicting entire cities, countries or even widely separated parts of the Earth in a matter of days or weeks. The Greek Historian Thucydides describes the plague of Athens of 429BC thus:

> "It is said to have begun in that part of Ethiopia above Egypt .... On the city of Athens it fell suddenly, and first attacked the men in Piraeus; so that it was even reported by them that the Spartans had thrown poison into the cisterns ...."

This event from Classical Greece bears striking similarities to the modern events relating to the present **COVID-19 pandemic**. Thucydides writes that many families were simultaneously struck by a disease with a combination of symptoms hitherto unknown. The idea of an enemy (the Spartans) poisoning the drinking water rings similar to what has happened in the Coronavirus outbreak in China and subsequent outbreaks taking place around the world.

The general belief, that is by no means well-proven, is that major pandemics, such as influenza as well as the present Coronavirus, start by random mutation or genetic recombination of a virus or bacterium which then spreads across a susceptible population solely by direct person-to-person contact. If this is so, it is somewhat surprising that major pandemics tend to be relatively short-lived, usually lasting about a year, and that they do not eventually affect the entire human population, which would not have a specific immunity of any totally new pathogen. We might argue that a primary cometary dust infection is potentially the most lethal, and that secondary person-to-person transmissions have *progressively reduced virulence*, resulting in a diminishing incidence of the disease over a limited period.

One important piece of historic evidence that emerged 101 years ago relates to the Influenza pandemic of 1918-1919 that caused some 20-30 million deaths worldwide. Reviewing all the available data, Dr. L. Weinstein wrote:

"Although person-to-person spread occurred in local areas, the disease appeared on the same day in widely separated parts of the world on the one hand, but on the other, took days to weeks to spread relatively short distances. It was detected in Boston and Bombay on the same day, but took three weeks before it reached New York City, despite the fact that there was considerable travel between the two cities. It was present for the first time at Joliet in the State of Illinois four weeks after it was first detected in Chicago, the distance between those areas being only 38 miles ...." L. Weinstein, *New England Journal of Medicine, May 1976.*

The lethal second wave of the Influenza pandemic of 1918 thus showing up on the same day in Boston and Bombay defies the realities of human travel at the time. This occurred before the advent of air travel, so it was impossible for people to transfer the virus from Boston to Bombay or vice versa. As Sherlock Holmes would have said: "The case is clear as daylight, my dear Watson: a new virus (or genetic trigger for a circulating virus) clearly fell through the skies simultaneously in locations that were separated by tens of thousands of kilometres."

COVID-19 Pandemic

In two publications in 2017 and 2019, written together with a team of colleagues, I had pointed out that the sun was approaching the deepest minimum for over a hundred years in its cycle of sunspots in late 2019 (See Figure 8). This, in turn, implies that the flow of high-speed electrons streaming out from the sun and which serves to maintain a protective sheath of magnetic field around the Earth is greatly reduced. With a weakened magnetic field, our planet would consequently be more "open" to the ingress of charged dust particles including bacteria and viruses from outside, and on this basis, we warned of the risk of impending pandemics.

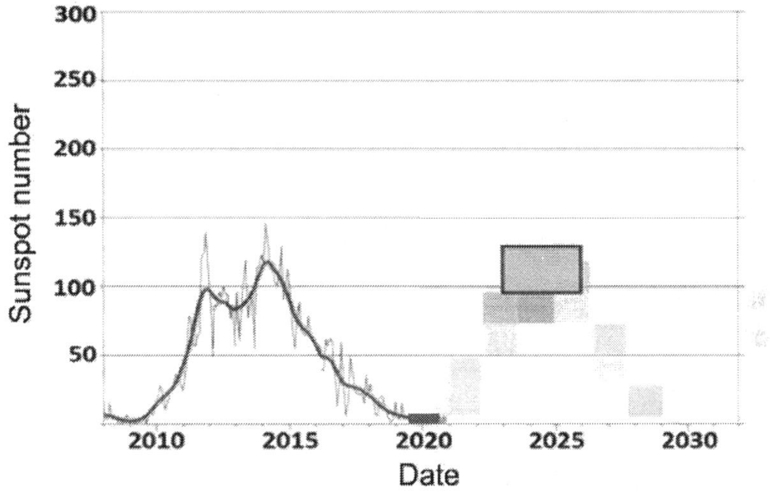

Figure 8: Graph presenting the sun approaching the deepest minimum in its cycle of sunspots in late 2019. (Wickramasinghe)

An indication of such an increased ingress of virus sized dust appears evident in the greatly enhanced frequency of noctilucent clouds that have been recorded throughout 2018 and 2019. Noctilucent clouds (NLC's) are tenuous cloud-like structures that show delicate filigree patterns visible after sunset against a not-yet-darkened evening sky. They are seen predominantly during the summer months in the latitude range 50° to 70°, both north and south of the equator, and are visible shortly after sunset during the period of astronomical twilight. In January 2020 noctilucent clouds were seen over Macquarie Island (9.8°S) and in March 2020 they were seen in the US as far south as Freedom, Oklahoma (+36°N) thus breaking all the normal rules of latitude limitation for the occurrence of noctilucent clouds (https://spaceweatherarchive.com/2020/03/26/noctilucent-clouds-over-the-south-pacific/). A dramatic picture of the distribution of NLC clouds on 12 June 2019 was captured by cameras onboard NASA's Aeronomy of Ice in the Mesosphere satellite (Figure 9). This image is a clear demonstration of the existence of a vast amount of comet/meteoroid dust of viral sizes at heights of 80km around which ice has condensed. If this included a cloud of viruses including the corona virus the first cases of 2019-nCoV infection in humans reported in China during November 2019 might be explained.

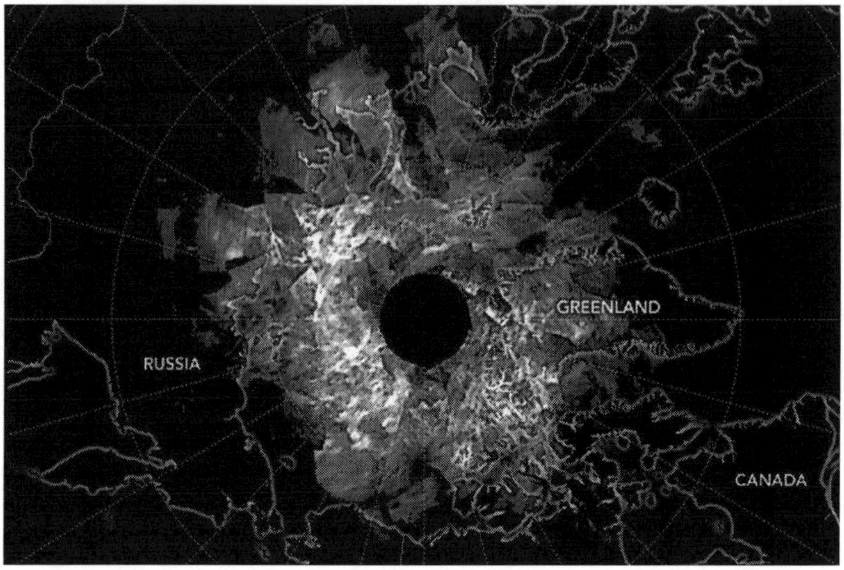

Figure 9: Visible light reflected from noctilucent clouds on 12 June 2019, taken by cameras aboard NASA's AIM satellite, showing noctilucent clouds stretching far south (*Photo Credit: Courtesy NASA*).

It is also interesting to note that this event followed remarkably close on the heels a cometary bolide that exploded on 11 October 2019 that lit up the midnight skies above north-east China.

My colleagues and I have discussed the possibility of the space origin of the Coronavirus, but the entrenched Earth-centred theory of transfers of a virus from an animal reservoir appears to prevail. My collaborator on this project Professor Edward Steele has analysed the gene sequences of the viruses from cases in Wuhan China and the USA and concluded that an animal transfer is ***not viable.*** So, if one is stuck with Earth centred theories, the conclusion is that the origin of the new virus is *unknown*.

Subsequent foci of COVID-19 outbreaks have developed along a narrow belt latitude 40-60 degrees N, with person-to-person transmission aggravating its spread within individual foci which seem to represent locations of new fall-out of the viruses. The spread of COVID-19 in the past months is looking similar in many respects to the patterns of spread seen during the 1918-1919 influenza pandemic. Let us hope that the reservoir of the COVID-19 causing virus from the stratosphere eventually becomes exhausted and together with the containment/social distancing measures in force the pandemic—like all pandemics—will come to an end.

Concluding remarks

When it is finally accepted that life is a truly cosmic phenomenon and that we are part of a vast cosmic biosphere, the implications for humanity will be profound. Even more important would be the acceptance that alien life in the form of microbes—bacteria and viruses—exist in our midst *even now* and continually rain down on our planet. Such microbes could be responsible for devastating pandemics, but more positively we should recognise cosmic viruses and bacteria could have the potential to augment our genomes—the genomes of all terrestrial lifeforms—and over long periods of time unravel an ever-changing panorama of life. The emerging facts pointing to the cosmic nature of life, when they are fully acknowledged, will mark an important turning point in human history.

The world is changing at an astonishing rate—a cliché, but true. The large subset of the changes that are distinctly for the worse are wrought by an insatiable desire to gain even greater control of the Earth's diminishing resources, and thereby to reassert our role as the supremely dominant species on the planet. In the quest for such dominance we are inevitably destroying the richness and diversity of life on Earth—plants, animals, microbiota—that has been established over millions of years.

Whilst advances in technology continue at accelerating pace, humanity as a whole is becoming ever more fractured. Wars and bitter sectarian conflicts and heart-rending suffering are to be seen everywhere. The "climate-change" marches and protestations of young people that are gaining momentum are perhaps emblematic of a desire to rebel against reigning paradigms that seem to be threatening our very existence.

Thomas Kuhn famously declared "... when paradigms change, the world changes with them." One could perhaps assert that a reversal of this causality is also possible—"when the world changes, paradigms can be forced to change."

Bibliography

1. Wickramasinghe, Chandra. 2015a. The Search for our Cosmic Ancestry (World Scientific Publishing, Singapore)

2. Wickramasinghe, Chandra. 2015b. A Journey with Fred Hoyle (World Scientific Publishing, Singapore)

3. Wickramasinghe, C., Wickramasinghe, K., Tokoro, G., 2019. Our Cosmic Ancestry in the Stars (Inner Traditions, NY)

4. Wickramsinghe, C., 2015. Where did we come from? Life of an astrobiologist (World Scientific Publishing, Singapore)

5. Reche, I, D'Orta, G., Mladenov, N. et al, 2018. Deposition rates of viruses and bacteria above the atmospheric boundary layer Deposition rates of viruses and bacteria above the atmospheric boundary layer, *The ISME Journal*, https://doi.org/10.1038/s41396-017-0042-4

6. Wickramasinghe, C. et al, 2019 Space weather and pandemic warnings? *Current Science, 117*, No.10

7. Harris, M.J., Wickramasinghe, N.C., Lloyd, D., Narlikar, J.V., Rajaratnam, P., Turner, M.P., Al-Mufti, S., Wallis, M.K. & Hoyle, F. (2001) The detection of living cells in stratospheric samples. Proc SPIE 4495, 192–198.

8. Shivaji, S., Chaturvedi, P., Begum, Z. et at, 2009. Janibacter hoylei sp. nov., Bacillus isronensis sp. nov. and Bacillus aryabhattai sp. nov., isolated from cryotubes used for collecting air from the upper atmosphere. *Int. J Systematics Evol. Microbiol*, 59, 2977–2986

Wickramasinghe et al., Astrobiol Outreach 2017, 5:2
DOI: 10.4172/2332-2519.1000159

Review Article

Sunspot Cycle Minima and Pandemics: The Case for Vigilance?
Wickramasinghe et al, Space weather and pandemic warnings?
CURRENT SCIENCE, VOL. 117, NO. 10, 25 NOVEMBER 2019

Q & A with Professor Chandra Wickramasinghe

By Melissa Layne

I would be remiss if I did not also include in our inaugural issue an interview with Chandra Wickramasinghe.

Question 1

Layne: Professor Wickramasinghe, thank you for the opportunity for this interview. Do you believe person-to-person viral transmission occurs with Coronavirus or other viruses? If the infection comes from space, how do you account for the "close quarters" effect where infection rates run so much higher on cruise ships and such?

Wickramasinghe: *The new virus can of course be infective from person to person, but it seem that it is infective only on relatively close contact. The large clusters of cases occurring simultaneously on cruise ships, or in isolated communities, can be easily explained if clouds carrying the infective virus comes down in local regions. As for freak "superspreaders" being responsible for large clusters of cases, I think this is a myth based purely on ignorance. If a group of people were exposed to a cloud of virus and became simultaneously infected from such an environmental source, there would of course be a dispersion, or spread, in the times when the illness shows up. That is to say the incubation period would have statistical spread, so one case will always appear to be the first. To crown him/her a title of superspreader, with mysterious power is akin to a medieval myth (explained in my article).*

These ideas were first discussed by the late Sir Fred Hoyle and me in two books—Diseases from Space (1979) and Evolution from Space (1980). Here, we introduced the theory that comets carry bacteria and viruses and that impacts by comets were important for both starting life on Earth and for its further evolution. The first point to make is that the standard view that life originates spontaneously on Earth in a primordial soup (or in deep-sea thermal vents) has no evidence whatsoever to support it. Every experiment that has been done to demonstrate or test this possibility has ended in dismal failure over more than 50 years. The molecular complexity of life – the information content of life – is of an exceedingly specific kind and it is super-as-

tronomical in quantity, and so the origin of life could not have happened on Earth. A few years ago the very oldest evidence of microbial life on Earth was discovered in rocks dated 4200 million years ago – and this was at a time when the Earth was being relentlessly pounded by comet and asteroid impacts. So there is little doubt now that the first life on Earth came with impacting comets, and the subsequent evolution of life happened against the backdrop of new bacteria and viruses being introduced via comets, adding new potential and new genes for evolution. It is this potential for evolution with new cosmic genes against which Darwinian evolution takes place. So there is no doubt cosmic viruses are in our genes. And this is the reason that new viruses coming from space today can relate to evolved life forms like ourselves.

Question 2

Layne: There is, to say the least, a lot of research and brain power being applied globally to Coronavirus. What are all those esteemed virologists missing in the data?

Wickramasinghe: *It is only relatively recently that scientists have been able to fully grasp the enormous magnitude of the microbial and viral content of the terrestrial biosphere. We now know that a typical liter of surface seawater contains at least 10 billion microbes as well as some 100 billion viruses—the vast majority of which remain unidentified and characterized to date. Two years ago an international group of scientists collected bacteria and viruses that fell through the rarefied atmosphere near the 4000 meter peaks of the Sierra Nevada mountains of Spain. They arrived at an astonishing tally of some 800 million viruses per square metre per day and an associated slightly smaller tally of bacteria – all of which would of course ultimately fall to the Earth's surface. The assumption normally made is that all such viruses and bacteria necessarily originate on the Earth's surface and are swept upwards in air currents. But in such a model many horrendous difficulties associated with the upward transport processes of bacteria and viruses are ignored. I think a significant fraction of this vast number of falling microbes must actually originate outside the terrestrial biosphere and come from cometary sources – viruses and bacteria that are expelled from comets.*

Further evidence comes from sampling the stratosphere for its bacterial and viral content. By sampling the stratosphere at a height of 41 km, using equipment carried using balloons already in 2002 we arrived at an estimated in-fall from this height of 20-200 million bacteria per square meter per day, and 10 to 100 times more viruses, falling downwards to the Earth. These are facts that cannot be ignored, but all too often they are! So, if we take into account all the facts available to date we cannot avoid the conclusion that vast numbers of bacteria and viruses continue to fall through the Earth's atmosphere, and it seems inevitable that a significant fraction is of external origin.

Question 3

Layne: How are comets and their debris stream meteor showers related to the hypothesis?

Wickramasinghe: *Of course comets have been regarded with fear and trepidation by many ancient cultures. Almost without exception comets have been thought to be bad omens – bringers of pestilence and death. The evidence for comets being implicated in the origin of life and also of diseases on Earth was intensely controversial when these ideas were first proposed in a scientific context by me in the 1970's. Things have only changed a little since these early days. Now there is a growing consensus that comets are in some way connected to the origin of life. But most people are still fearful about going any further. But the facts alone tell us a different story. If one looks at all the available facts on epidemics throughout history, comets and epidemics appear to be causally linked.*

Stories of the sudden spread of plagues and pestilences punctuate human history throughout the millennia. The various epidemics, scattered through history and throughout the world often bear little or no resemblance one to another. But they share a common feature. They often affect entire cities, countries or even widely separated parts of the Earth in a matter of days or weeks.

The Greek Historian Thucydides describes one such epidemic—the plague of Athens of 429 BC thus:

> *"It is said to have begun in that part of Ethiopia above Egypt .... On the city of Athens it fell suddenly, and first attacked the men in Piraeus; so that it was even reported by them that the Peloponnesians had thrown poison into the cisterns ..."*

This event from Classical Greece bears striking similarities to the modern events relating to the outbreak of the Coronavirus in China. Thucydides writes that many families were simultaneously struck by a disease with a combination of symptoms hitherto unknown. The idea of an enemy (the Spartans) poisoning the drinking water rings true to what has happened in the Corona virus outbreak in China.

Very similar descriptions of a sudden onset and rapid global spread is relevant to almost all earlier as well as later epidemics. Extreme swiftness of transmission is hard to comprehend if, as is usually supposed, infection can pass only slowly from person to person or be carried by vectors such as lice and ticks, and more recently, monkeys, bats or snakes. Such explanations are particularly untenable for the many epidemics that occurred before the advent of air travel when movement of people across the Earth was a slow and tedious process.

The general belief, that is by no means well-proven, is that major pandemics, such as influenza, start by random mutation or genetic recombination of a virus or bac-

terium which then spreads across a susceptible population by direct person-to-person contact. If this is so, it is somewhat surprising that major pandemics tend to be relatively short-lived, usually lasting about a year, and that they do not eventually affect the entire human population, which would not have a specific immunity of any totally new pathogen. We might argue that a primary cometary dust infection is potentially the most lethal, and that secondary person-to-person transmissions have progressively reduced virulence resulting in a diminishing incidence of the disease over a limited period. Infections of a human population could occur directly by contact with "infected" meteoritic dust from an exploding cometary bolide, or indirectly by the original cometary infection passing first to rats, lice, primates, bats, snakes which can act as intermediaries.

One important piece of historic evidence that emerged 101 years ago relates to the great Influenza pandemic of 1918-1919 that caused some 20-30 million deaths worldwide.

Reviewing all the available data Dr. L. Weinstein wrote as follows:

> *"Although person-to-person spread occurred in local areas, the disease appeared on the same day in widely separated parts of the world on the one hand, but on the other, took days to weeks to spread relatively short distances. It was detected in Boston and Bombay on the same day, but took three weeks before it reached New York City, despite the fact that there was considerable travel between the two cities. It was present for the first time at Joliet in the State of Illinois four weeks after it was first detected in Chicago, the distance between those areas being only 38 miles ..." L. Weinstein, New Eng.J.Med, May 1976*

The lethal second wave of the influenza pandemic of 1918 thus showing up on the same day in Boston and Bombay defies the realities of human travel at the time. Before the advent of air travel it was impossible for people to transfer the virus from Boston to Bombay or vice versa. As Sherlock Holmes would have said: "The case is clear as daylight, my dear Watson: a new virus (or genetic trigger for a circulating virus) clearly fell through the skies simultaneously in locations that were separated by tens of thousands of kilometres."

Over the following 12 months the infective agent probably became dispersed through the troposphere and came down with an expected seasonal modulation across much of the world.

The abrupt appearance in the literature of references to particular diseases is also significant to recall in that they probably indicate specific invasions of new pathogens. Thus the first clear description of a disease resembling influenza was probably recorded in the 17th century AD, while the earliest reference to the common cold in the literature was about the 15th century AD. Also, it is significant that many historic plagues such as the Plague of Athens (described in meticulous detail by Thucydides) have not been linked to easily recognisable modern counterparts.

Question 4

Layne: Could you summarize the most important facts with regard to the COVID-19 outbreak and how they relate to your long-standing theory of disease from space?

Wickramasinghe: *On October 11 2019 a meteoritic bolide (probably the fragment of a comet) exploded in a brief flash in North East China. I think it probable that a piece of this bolide containing trillions of the COVID-19 virus broke off from the bolide as it was entering the stratosphere (above 50km), and became dispersed through the troposphere. A large of this fell first on Wuhan, the epicentre of the virus outbreak causing a huge number of cases and deaths on a very short timescale. New independent foci of the disease appeared later in North Korea, Iran, Italy, Spain, California, New York and other locations all lying remarkably within a narrow belt of latitude 30-50 degrees North. I think the virus cloud was probably carried by atmospheric currents in a jet stream and came down in various locations in this belt of latitude. Over 98 percent of all the cases, and deaths, lie in countries located in this belt. Once the virus cloud falls to ground level then of course person to person spread takes over.*

My colleague Professor Ted Steele, who is a distinguished immunologist and biomedical scientist, has shown that isolates of the corona virus taken from the different foci of disease show almost no mutations. This indicates that the incoming virus was essentially a "monoculture". This is of course dramatically different to the picture one gets if the main spread of the virus was through affected victims replicating the virus and distributing copies which inevitably would show mutations over a broad sample of isolates – certainly from one focus of the disease to the next. Everyone in the Wuhan region would have been exposed to essentially the identical virus (including many animals, such as mammals, snakes and even perhaps vegetation). The same would be true for the other foci of the pandemic. And, by the way, Steele has shown convincingly that the idea that the virus originated from bats, via pengolins and then hopped to humans is utterly impossible. This is on the basis of the genome sequencing studies that he has done over the past several weeks.

Claims that people can pass on the virus to others without showing any symptoms appears somewhat strange, and perhaps far-fetched. On the other hand, the cometary hypothesis is consistent with a wide regional "environmental" contamination following a fallout of the virus - that may include contamination of clothes, hair, cars, sidewalks, trees, grassland, surface water pools and water reservoirs. It is of course evident that some degree of human-to-human transmission of the virus must occur particularly in situations of close contact. This is what the present lock-down and social distancing measures in many countries might aim to restrict. But the main causes are not addressed.

I think it is significant that passengers on cruise ships have also been affected without significant evidence of intimate contact with infected individuals. A very similar phenomenon was also noticed during the 1918-1919 influenza pandemic when outbreaks were recorded in groups of people who were essentially in isolation at sea over many weeks.

The factors governing the actual pattern of global incidence for any particular extra-terrestrial pathogenic invasion could be complex, depending on local meteorology and the seasons. If bacteria or viruses are dispersed in a diffuse cloud of small particles throughout the atmosphere, the incidence of disease may well be global. On the other hand, a smaller disintegrating aggregate of infective clumps falling over a limited area of the Earth's surface could provide a geographically more localised outbreak of disease.

In particular certain latitude belts might well be more favoured than others for either the accumulation or the settling of these particles - or indeed for their avoidance. High mountain ranges such as the Himalayas and the Alps that puncture the troposphere would be ideal locations that act like "bath plugs" for draining clouds of particles that are spread around the world. So also would arctic regions, where the troposphere is particularly thin (6-7 km) during the winter. It would not be surprising to find a surge of cases of COVID-19 reported in the future in any of these geographical locations. Indeed, during the 1918-1919 influenza pandemic, arctic communities in Alaska, far removed from major population centres, were "mysteriously" struck. This is not unlike some current situations reported for COVID-19 where no epidemiologic link could be traced to distant centres of infection.

Question 5

Layne: Professor Wickramasinghe, I have noticed many nations are spraying disinfectant in public areas, but not the US. Does your comment above suggest the United State should spray disinfectant in order to remove the virus from public spaces? I thought Coronavirus did not live long on surfaces, or is still falling in?

Wickramasinghe: *I think it is vitally important that we approach such questions with humility. There remains much about this virus—its origin and modes of spread that is very poorly understood. The prevailing confidence of "expert" opinions stems firstly from the assumption that the virus is known to have originated via animals, or perhaps a sequence of animals. This as I said is an idea that seems to be deeply flawed; and if this one assumption turns out to be wrong all that rests upon it is open to serious question. The alternative view that scientists have been brainwashed to disregard as being crazy and inadmissible is that it came from space. Following the work that was started in the late 1970's in collaboration with the late Sir Fred Hoyle I have now examined the epidemiology of many epidemics of respiratory viral diseases over several decades and presented a cogent case to say new viruses, in many instances, have an extraterrestrial origin. This may sound crazy only because we have been locked into a paradigm of Earth-centred biology and Earth-centred evolution. There* *is now growing evidence for life—its origin and all its evolutionary history—being inextricably linked to the universe at large.*

A major paradigm shift is now staring us in the face, and we are still trying desperately to resist it at all cost. If the Earth's biosphere extends to infinity—as is now absolutely clear—the concept of COVID and other viruses coming from space can no longer to be considered impossible or outrageous in any way.

I believe there are good arguments for asserting that the present pandemic was a combination of several massive infall events (infall of COVID-19 virus laden dust) over the general latitude belt 40-60 degrees north followed by contagion and person to person spread. We have argued that the sudden onslaught starting in China and moving to South Korea, Iran, Italy looks very much like such events. In these cases, one has an instinct to think that surface cleaning will have an effect. The Chinese were seen disinfecting road surfaces, exterior of buildings—everywhere—with high speed jets of disinfectant. This may well have worked in their favour.

Question 6

Layne: Is there reason to believe that attention and data gathering, and subsequent epidemiological follow-up studies, may reveal flaws in the current paradigm so egregious that the scientific thinking is forced to shift to disease from space?

Wickramasinghe: *I believe that in the fullness of time this will be so—it must be so. Neglecting to admit our biological connection with the external universe could be a matter of life and death for humanity. We have stated many times that the technology for monitoring the stratosphere for incoming viruses does indeed exist at the present time and moreover is amazingly inexpensive. Having a stratospheric surveillance programme in place could forewarn us of future pandemics before they hit the ground and hit us hard. A paradigm shift towards admitting our cosmic connection and cosmic destiny is in my view long overdue. I believe that in the long term the truth always prevails.*

Question 7

Layne: Please describe what you believe to be the most likely general scenario for the manner in which viral particles survive the entry into earth's atmosphere. My understanding is that you believe viruses are orders of magnitude smaller than the tiny bits of comet they are encapsulated in, does that in some manner provide them protection from high speed atmospheric conditions? If they can't survive soap, how do they survive atmospheric entry?

Wickramasinghe: *The entry of microbes and viruses is discussed in an appendix to our 1979 book Diseases from Space, that is currently being reprinted. While large* *in-falling meteoroids and millimetre-sized fragments get heated as they enter the atmosphere at speeds of 10km/s, and many burn as meteors, very much smaller structures do indeed survive. For example, an atom or molecule survives and even a virus coated with a ultrating coating survives high speed entry. Also, larger centimetre sized fluffy, highly porous particles can survive entry because they have vibrational modes that get rid of the energy of impacting atoms. In fact clumps of viruses have been stuck onto the outside of a rocket and found to survive as it was shot through the atmosphere.*

I cannot comment on the action of soap on viruses, but I suppose the soap can act chemically with the virus and perhaps inactivate it.

Question 8

Layne: In terms of space education, are there universities that have integrated your work into their curricula?

Wickramasinghe: *My 1960's discovery that cosmic dust is carbonaceous and organic was bitterly contested at the time but now is coming to be taught in mainstream astronomy, although my clear and unambiguous original priority is often forgotten. Then when in 1981 Sir Fred Hoyle and I proposed that astronomy and biology should come together, we were ignored at the time, but by the year 2000 astrobiology was recognised as a new subject as is taught in many universities. Again we are given very little credit. But this is an old story. The Pre-Socratic Greek philosopher Anaxagoras (c. 500—428 BCE), who declared that the Sun and Moon were physical objects not gods or goddesses, was banished from Athens. But he was of course vindicated and honoured in the fullness of time.*

Question 9

Layne: Are there research opportunities—whether with NASA or SpaceX space missions, to collect more evidence to support Panspermia? (Confirmation of Microbial Ingress from Space) Are your colleagues continuing to advance your work?

Wickramasinghe: *My colleagues and I have been trying our best to alert the scientific community, and trying to spur them into action—simply to accept the most straightforward of facts. So far, we have not succeeded, although evidence supporting panspermia and our cosmic heritiage continues to come in fast and thick. The persistence with which major establishments turn away from facts will go down in history as the biggest societal scandal of our age. If they accept the evidence dispassionately, new vistas of science will open up before us, which it would be the privilege of future generations to explore. If they do not, in the words of Bertand Russell, "nothing lies before us but universal death".*

Question 10

Layne: My university, American Public University System, has a large military student population and we would like to develop a program for students that includes learning about space-related advancements (i.e., astrobiology) alongside various military applications aimed toward homeland security and the U.S. Space Force. I believe that your work should be integrated in this type of program—especially when you mentioned the need for a stratospheric surveillance programme in place that could forewarn us of future pandemics. In what other circumstances do you see an immediate need for this type of program? How would you structure it?

Wickramasinghe: *I hold, amongst other unpaid positions, an Honorary Professorship at the Sir John Kotelawala Defence University in Sri Lanka! I think the most important activity that is neglected by all universities and governments is to track the infall of bacteria and viruses from the skies—from space. We have clearly shown over the past several years that an infall of tons of bacteria descends from space every year, and so we have urged governments to monitor the skies for our safety. Earth-centred biological thinking has so far prevented this from happening but I hope this will change after our recent experience with the COVID-19 pandemic. My friend Arthur C. Clarke once told me that the dinosaurs became extinct because they did not have a space guard programme! Let's hope that does not happen to us humans, because we are so incredibly stupid.* Ω

Personal website: www.profchandra.org

Buckingham astrobiology website: http://www.buckingham.ac.uk/research/bcab

Ruhuna astrobiology website: http://physics.ruh.ac.lk/astrobiology/

ISPA website: www.ispajapan.com

References

Hoyle, F., & Wickramasinghe, C. (1979). Diseases from space. New York.: Harper & Row.

Steele, E. J., Al-Mufti, S., Augustyn, K. A., Chandrajith, R., Coghlan, J. P., Coulson, S. G., ... & Louis, G. (2018). Cause of Cambrian explosion-terrestrial or cosmic? *Progress in Biophysics and Molecular Biology, 136*, 3-23.

Hoyle, F., & Wickramasinghe, N. C. (1981). Evolution from space. JM Dent.

Hoyle, F., & Wickramasinghe, N. C. (1986). The case for life as a cosmic phenomenon. *Nature, 322*(6079), 509-511.

Hoyle, F., & Wickramasinghe, C. (1993). Our place in the Cosmos: the unfinished revolution. JM Dent.

Morrison, David (January–February 2007). "Carl Sagan's Life and Legacy as Scientist, Teacher, and Skeptic". *Skeptical Inquirer. 31* (1): 29–38. ISSN 0194-6730. Archived from the original on February 1, 2016.

Wickramasinghe, N. C. (1974). Formaldehyde polymers in interstellar space. *Nature, 252*(5483), 462-463.

Wickramasinghe, C. (2015). The search for our cosmic ancestry.

Wickramasinghe, C. (Ed.). (2015). Vindication Of Cosmic Biology: Tribute To Sir Fred Hoyle (1915-2001). *World Scientific,* 2018.

Books

Interstellar Grains: Chapman & Hall, London, 1967

Light Scattering Functions for Small Particles with Applications in Astronomy: J. Wiley, 1973

Solid-State Astrophysics: (ed with D.J. Morgan) D. Reidel Co., 1975

Interstellar Matter: (with F.D. Khan & P.G. Mezger) Swiss Astron.Soc., 1974

Cosmic Laboratory: University College, Cardiff Press, 1975

Lifecloud: The Origin of Life in the Galaxy: (with Fred Hoyle) J.M. Dent, Lond., 1978

Diseases from Space: (with Fred Hoyle) J.M. Dent, Lond., 1979

Origin of Life: (with Fred Hoyle) University College Cardiff Press, 1979

Space Travellers: The Bringers of Life: (with Fred Hoyle) University College Cardiff Press, 1981

Evolution from Space: (with Fred Hoyle), J.M. Dent, 1981

Is Life an Astronomical Phenomenon?: University College, Cardiff Press, 1982

Why Neo Darwinism does not Work: (with Fred Hoyle) University College Cardiff Press, 1982

Proofs that Life is Cosmic: (with Fred Hoyle) Inst. of Fund.Studies, Sri Lanka, Mem, No. 1, 1982

From Grains to Bacteria: (with Fred Hoyle) University College, Cardiff Press, 1984)

Fundamental Studies and the Future of Science: (editor) University College Cardiff Press, 1984

Living Comets: (with Fred Hoyle) University College, Cardiff Press, 1985

Viruses from Space: (with F. Hoyle a nd J. Watkins), University College Cardiff Press, 1986

Archaeopteryx - The Primordial Bird: A Case of Fossil Forgery: (with Fred Hoyle) Christopher Davies, Swansea, 1986

Cosmic Life Force: (with Fred Hoyle), J.M. Dent, Lond., 1988

The Theory of Cosmic Grains: (with F. Hoyle), Kluwer Academic Publishers, 1990

Our Place in the Cosmos: (with Fred Hoyle) Weidenfeld and Nicholson, Lond., 1993

The Wonders of Life and the Universe (with Daisaku Ikeda) Mainichi Press, 1992, 1993

Glimpses of Life, Time and Space (a book of poems) Writers' Workshop, Redbird, 1994

Life of Mars: The Case for a Cosmic Heritage: (with Fred Hoyle) Clinical Press, 1997

Space and Eternal Life (a dialogue with Daisaku Ikeda) Journeyman Press, 1997

Astronomical Origins of Life: Steps towards panspermia (with F. Hoyle) Kluwer Academic Press, 2000

Cosmic Dragons: Life and Death on Our Planet. Souvenir Press, 2001

Fred Hoyle's Universe (eds with G. Burbidge and J.V. Narlikar) Kluwer Academic Publ. 2003

A Journey with Fred Hoyle: The search for cosmic life, World Scientific and Imperial College Press, 2004

Comets and the Origin of Life (with J.T. Wickramasinghe and W.M. Napier) World Scientific and Imperial College Press, 2009

The Search for Our Cosmic Ancestry, World Scientific, 2015

Where did We Come From? World Scientific, 2015

Vindication of Cosmic Biology: Tribute to Sir Fred Hoyle, World Scientific, 2015

The Big Bang and God: an Astrotheology (with Theodore Walker, Jr) Palgrave Macmillan, 2015

The Search for Our Cosmic Ancestry, World Scientific, 2015

A Destiny of Cosmic Life—Chapters in the life of an astrobiologist, World Scientific, 2016

Where did we come from?, World Scientific, 2015

Proofs that Life is Cosmic, World Scientific, 2017

Cosmic Womb (with Robert Bauval), Inner Traditions Inc., 2018

Our Cosmic Ancestry in the Stars (with K. Wickramasinghe and G. Tokoro), Bear and Co. Rochester, Vermont, USA.

Images Capturing the Historic SpaceX Demo-2 Launch

Melissa Layne
American Public University System

"A picture is worth a thousand words." This English adage could not ring more true than during the recent SpaceX NASA Demo-2 launch on May 30th, 2020. How so? This launch was markedly different from other historic launches—in many aspects. One stark difference, is that the American public was strongly discouraged to physically attend the event due to potential spread of COVID-19. The COVID-free environment of yesteryear allowed Americans to physically gather and share the excitement, pride, and enthusiasm of such a momentous event. When NASA's Commander Doug Hurley flew on NASA's last *Atlantis* mission in 2012, he shared his thoughts on physically attending a launch:

> "Until you see one in person, you really haven't seen a shuttle launch. It really is an emotional experience to actually see the boosters light and see the shuttle head skyward as it starts to catch the space station. I want as many folks as possible to see a shuttle launch and realize what this country has accomplished."
>
> ~Commander Doug Hurley

Fellow *Atlantis* NASA astronaut, Rex Walheim agreed,

> "When you get to see a launch, it's not just witnessing a launch, you experience it, It's something you hear, you feel, you experience. I think people will be proud to be part of a country that can take this magnificent vehicle and sling it into orbit, and just see the incredible power and majesty of this vehicle taking off. It's a real treat."
>
> ~Rex Walheim

Although the public did not have the opportunity to physically attend and encounter all of the sensorial experiences of the SpaceX Demo-2 launch live, the launch was scheduled at a moment in time where Americans, and even those in other countries, needed a sense of enlightenment, identity, solidarity, and renewed strength. In different ways, this historic event gave many of those things back to us. We watched the launch live on NASA TV and various other social media outlets on our televisions, computers, laptops, and smartphones. As we watched the smoke slowly simmering around Falcon's thrusters and counted down the final seconds until lift-off, Crew Dragon lifted from the ground, and, if only for a brief moment in time, seemed to take with it the crippling weight of our world and up into space.

Returning back to the adage, "a picture is worth a thousand words"; sometimes the conveyance of meaning or essence of powerful experiences can be expressed effectively by a simple, still image, rather than trying to provide a verbal description. The following pages present a culmination of images, that speak volumes.

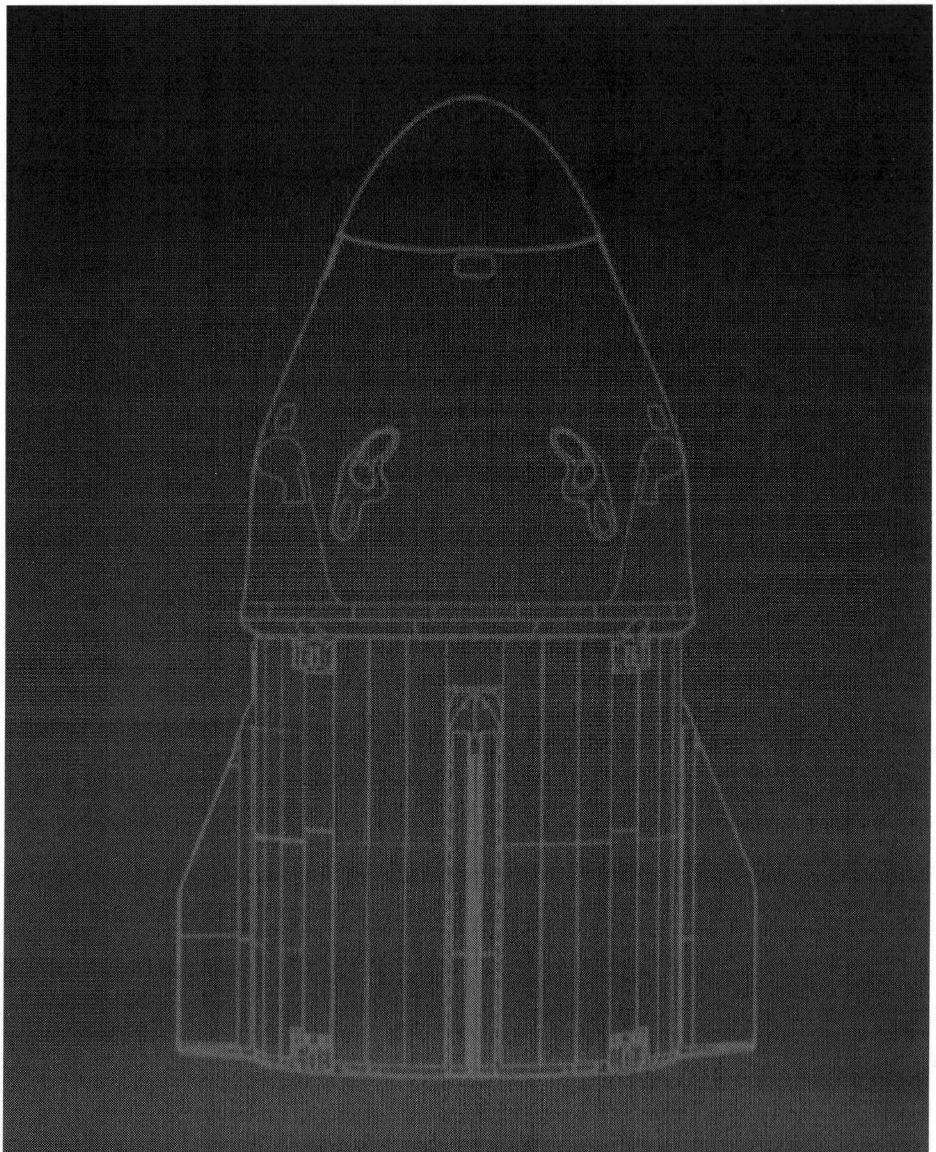

The blueprint above presents the Crew Dragon, which has the cabability of carrying up to 7 passengers into orbit, and large amounts of cargo. Even more impressive, it is the first ever commercial spacecraft to carry humans to the ISS. *Courtesy SpaceX*

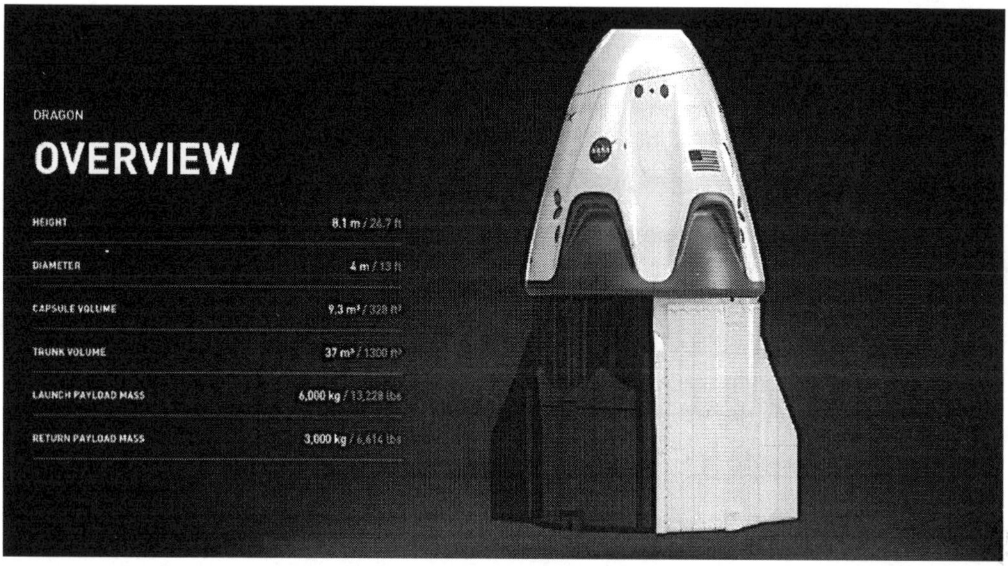

Dragon's specifications and a view of the interior seating inside of the Crew Dragon capsule. *Courtesy SpaceX*

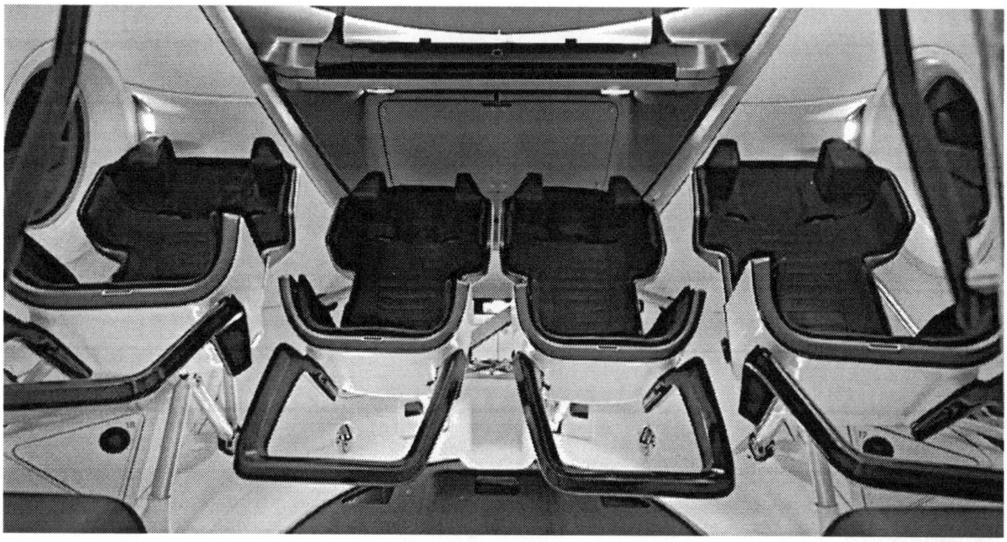

(Top) Crew Dragon resting in the Kennedy Space Center hangar. (Middle left) Ready for take-off on May 30th 2020. (Middle right) Behnken and Hurley enter the spacecraft via an access arm walkway. (Bottom) A side view of the walkway. *Courtesy SpaceX*

Designed by SpaceX, Behnken and Hurley traveled to space in style.
Courtesy SpaceX

(Top) Behnken and Hurley get their space helmets properly secured. (Bottom) Doug Hurley (second from right) travelled on the last NASA mission, Atlantis back in 2011. Also pictured, Commander Chris Ferguson, Sandy Magnus and Rex Walheim. *Courtesy Reuters*

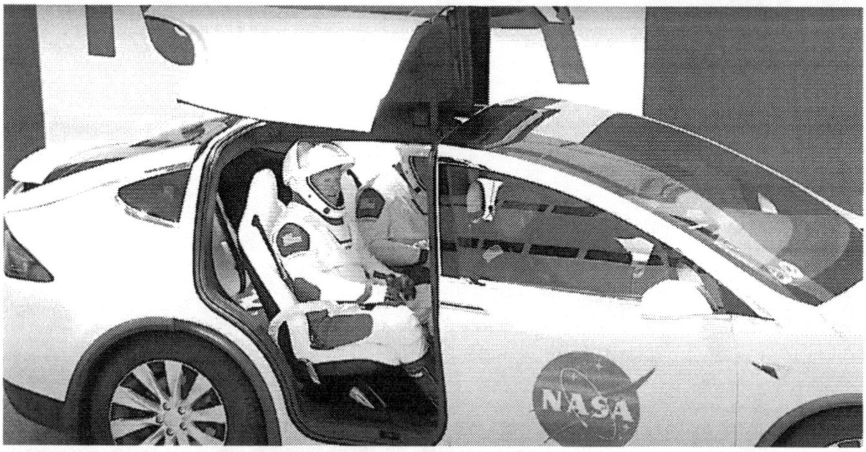

(Top) NASA Administrator, Jim Bridenstine, captures a quick "socially distant" selfie with both astronauts before launch. (Middle and bottom) Behnken and Hurley are escorted to the launch pad in their Tesla Model X sport cars. *NASA TV/UPI*

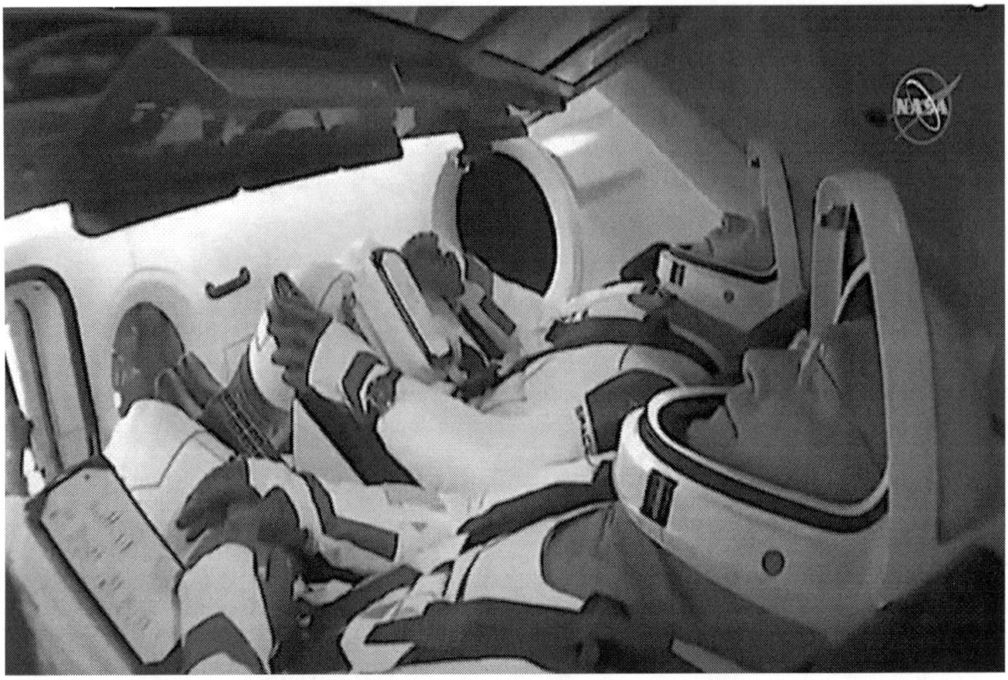

(Top) Behnken and Hurley securely situated in their Crew Dragon capsule custom seats. (Bottom) Ground control at Kennedy Space Center. *Photo by SpaceX/UPI*

Preparing for lift-off. *Courtesy SpaceX*

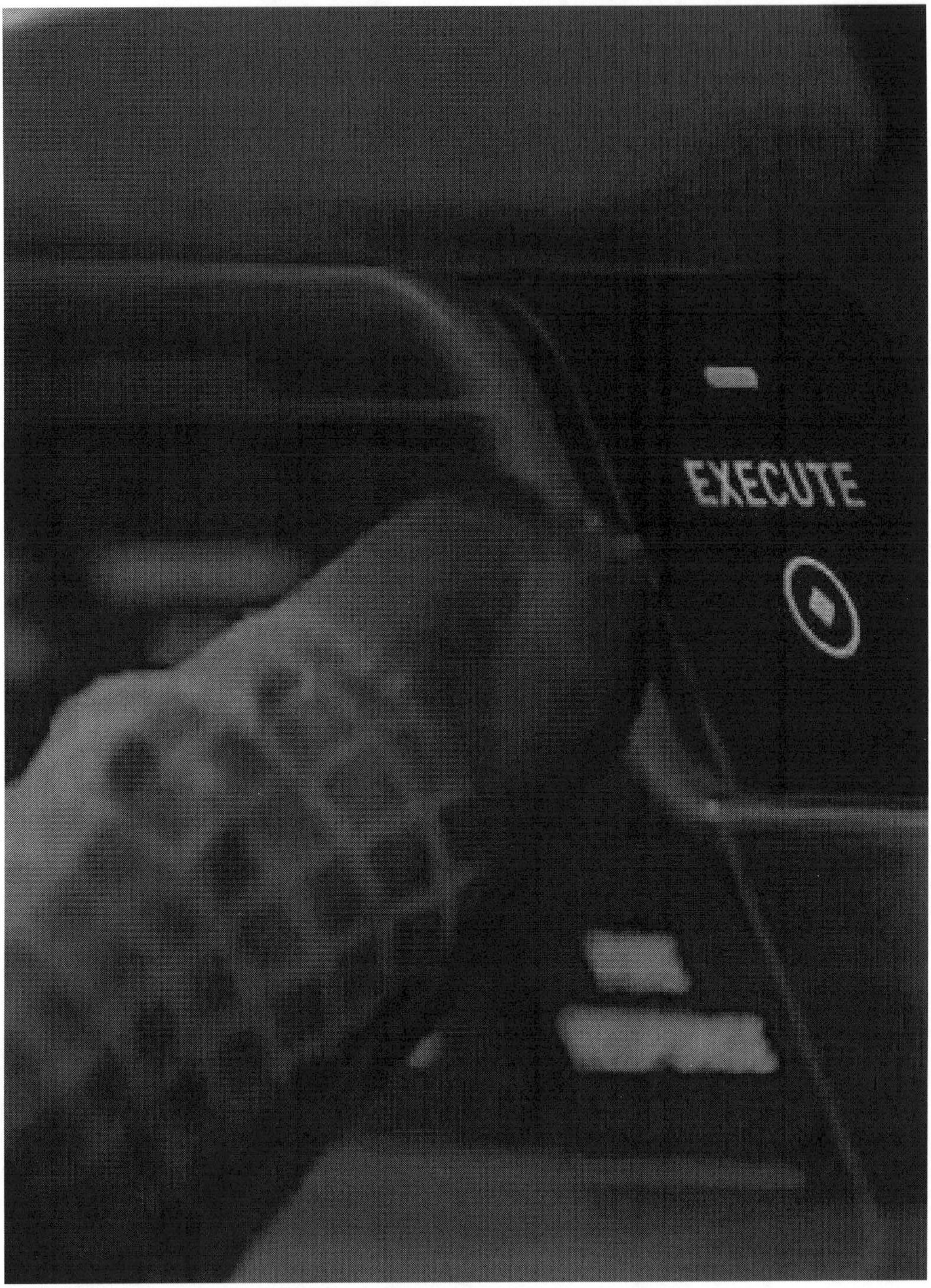

Three…Two…One. *Courtesy SpaceX*

Images of the successful Demo-2 launch. *Courtesy SpaceX and NASA*

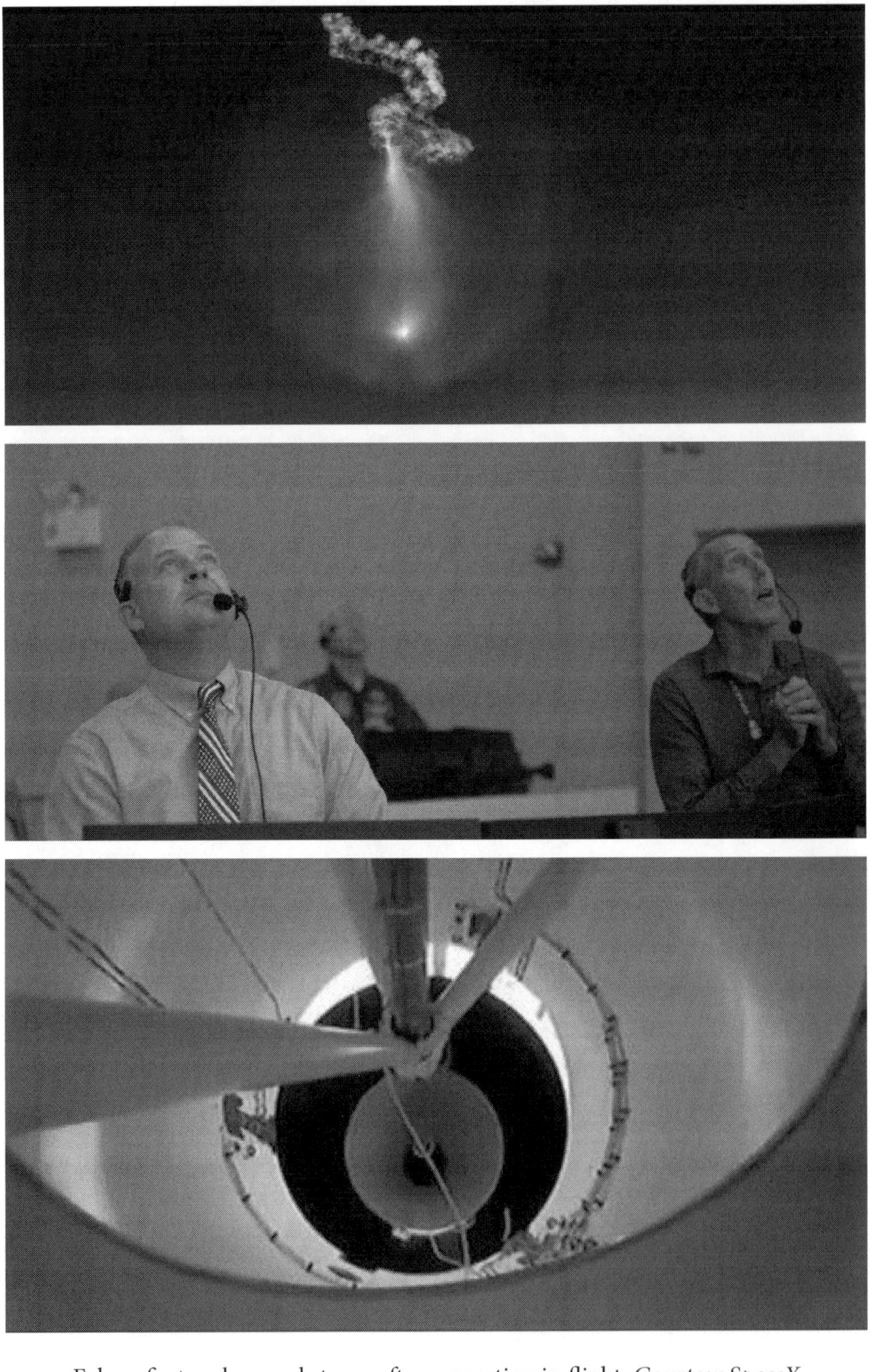

Falcon first and second stages after separating in flight. *Courtesy SpaceX*

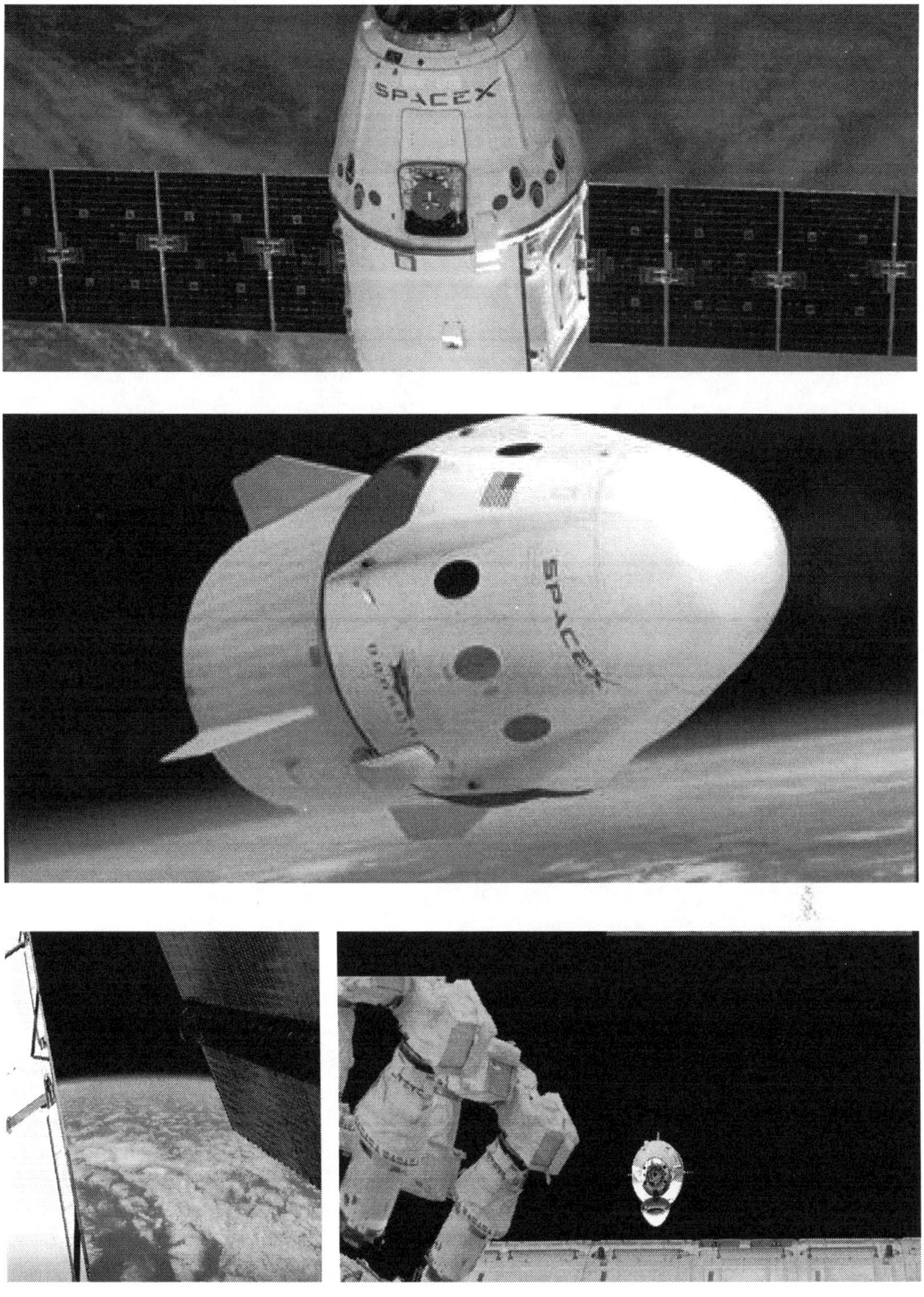

Crew Dragon capsule in orbit. *Photos courtesy of NASA/UPI and SpaceX*

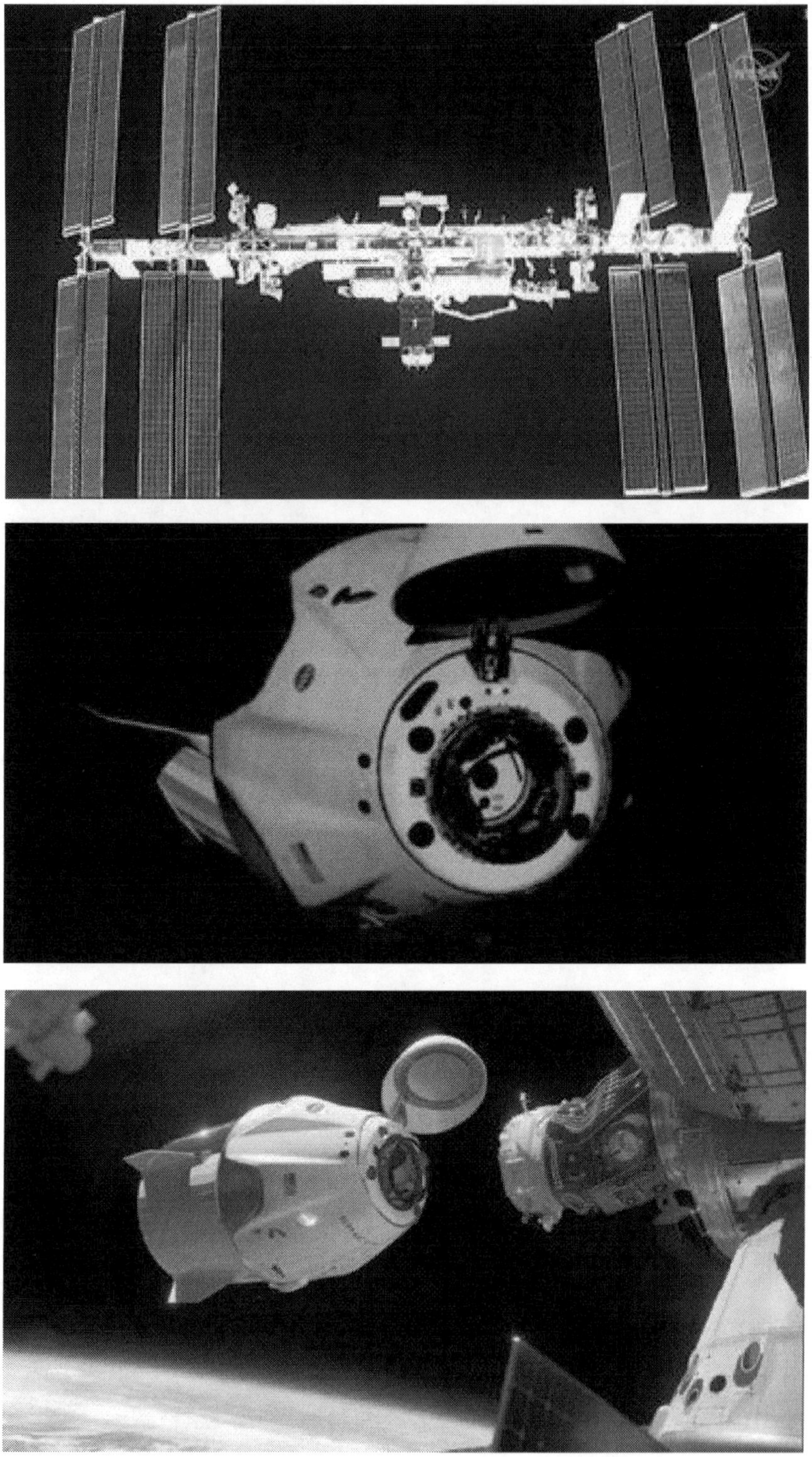

(Top) International Space Station. (Middle) Crew Dragon approaching docking station. (Bottom) Crew Dragon capsule preparing to dock. *Photos courtesy of NASA and SpaceX*

President Donald Trump, First Lady Melania Trump, and Vice President Mike Pence attending the launch.

SpaceX CEO, Elon Musk celebrating the historic, successful launch

American Public University System Observatory

Ed Albin

American Public University System

The American Public University System observatory is located on the main center campus in Charles Town, West Virginia, standing atop the tallest building on the site with a bright, silver 22 and a half feet wide dome. Within its protective cavern the CDK 24 inch diameter telescope is housed, mounted on a PlaneWave, A200 equatorial pier that stands at an overall height of 10 feet. There are not one but two telescopes attached to the central pier, with a 5 inch diameter Tele Vue refractor fastened on top of the 24 inch using additional brackets. Both telescopes use an SBIG type of CCD camera, the CDK uses a model ST 16803 and the Tele Vue uses an ST 8300, each with full set filter wheels for different research and photographic needs. The power of this set up lies in its equipment as much as its remote operations capabilities, wherein faculty, professors and even students may access and control the telescopes from vastly different and far away locations in pursuit of any astronomical objects.

There are several objects that have been or are continuing to be observed and projects carried out with the APUS observatory, including variable stars, nebulae, galaxies, exo-planets, supernovae, and star clusters. Some of the research conducted has been part of an ongoing effort, like the supernova search project, led by Dr. Kristen Miller, where weekly or biweekly images are obtained of over 140 different galaxies and made easily accessible to volunteering students to look over as they analyze them for any newly occurring transients as possible supernovae. There are 18 volunteers actively searching these images and reporting their results, which allows for open access to real data for the students and a way to enable greater learning and science. Another project involved taking photometric images of several variable stars and potential variable stars as highlighted by the American Association of Variable Star Observers (AAVSO). This project involved taking images and analyzing them using open software hosted by the AAVSO as well as installed on the APUS observatory suites to measure star magnitudes and report them to the variable star community. The resultant light curves is currently being analyzed and used for a paper in the APUS SPST690 (Independent Study) class, and the data is now readily available for other observers to work with or include in future research. The observatory has also participated in efforts to track and detect new asteroids, which is an ever-growing part of the space science fields, for both earth security measures and increasing science goals of understanding the ancient building blocks of the solar system. Other such solar systems are also of interest by the faculty and staff at APUS, as they use the observatory to image known stars hosting exoplanets and add transit data to the scientific community for continuing research.

Furthermore, beautiful photographic works have been undertaken at APUS using the CCD cameras and installed software to image and process several nebulae, galaxies and more. A listing and brief description of some color processed images composed with the APUS CDK 24-inch telescope can be found below.

The Horsehead Nebula in the constellation Orion is a dark and emission nebula dominated by hydrogen gases. This shot was taken with the combination of several R, G, and B filtered images each taken at 180 seconds and stacked together.

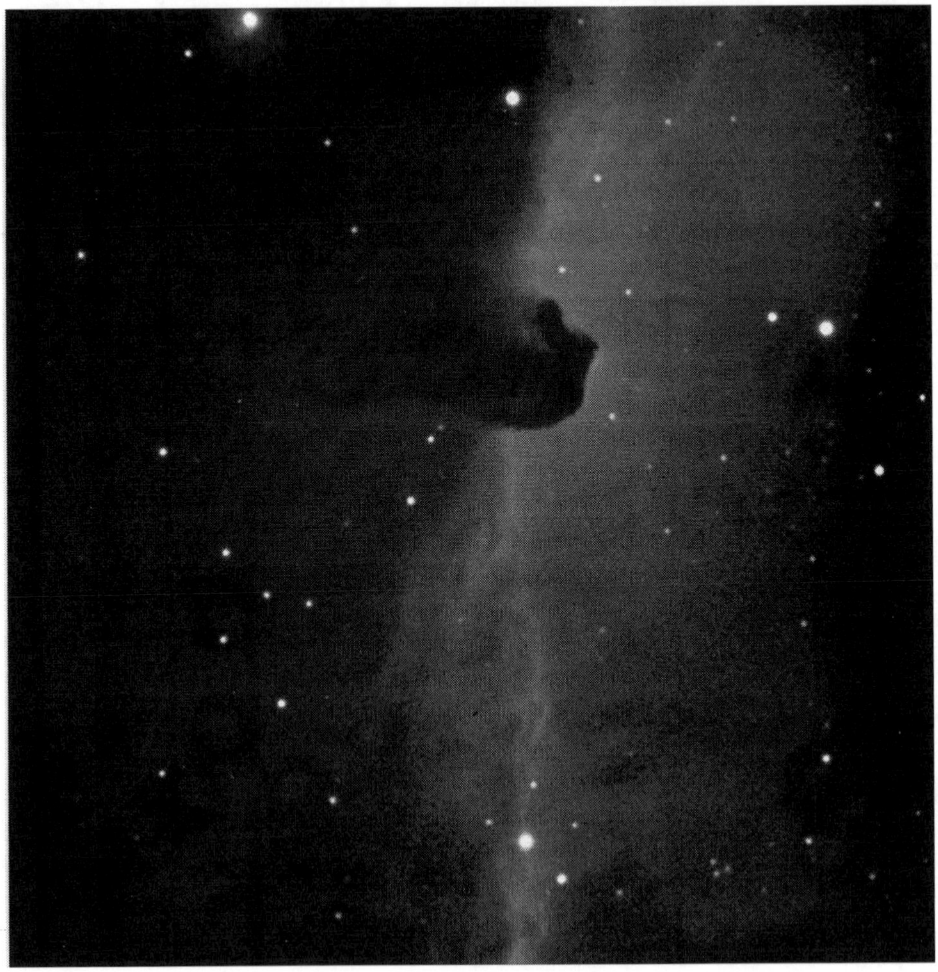

Figure 1. The Horsehead Nebula

The Rosette Nebula is a vast nebula that only partially fit in the field of view of the CDK 24-inch telescope. Nevertheless, its rendering is seen here, also taken with the R, G and B filters and 180-second exposures and later stacked.

Figure 2. The Rosetta Nebula taken with the R, G and B filters and 180-sec exposures

Galaxy M81 or Bode's Nebula, was imaged in the constellation Ursa Major and was cataloged by Charles Messier in August, 1779. This image is an LRGB composite of 180-second exposures.

Figure 3. Galaxy M81 or Bode's Nebula

Similarly, Galaxy M82 is seen here and is just outside the field of view when centrally focused on M81 by the CDK 24-inch telescope. This galaxy is known as the "Cigar Galaxy" and was also discovered by Messier in the late summer of 1779.

Figure 4. Galaxy M82, also known as the "Cigar Galaxy"

The Wizard Nebula, also known as NGC 7380 was one of the first color images composed on the APUS observatory from the CDK 24 inch telescope. It took R, G and B color filters at 180-second exposures.

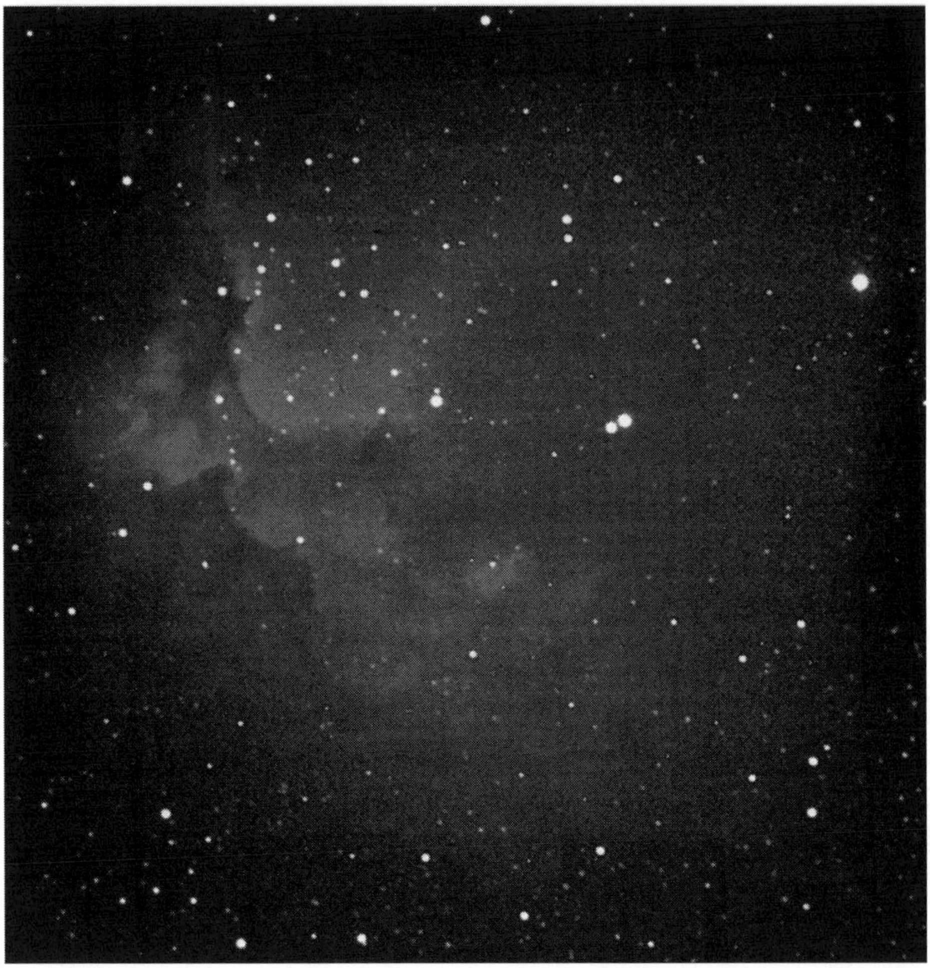

Figure 5. The Wizard Nebula, also known as NGC 7380

Finally, the emission nebula NGC 281 using the LRGB color filter imaging suite was used in the stacked resultant pictures seen below.

Figure 6. Emission Nebula NGC 281

The APUS observatory has become a great resource for the university's research capabilities and continuing efforts to expand the space studies program to better meet the needs of students, faculty and the science community as a whole. Situated higher than its surroundings, the observatory markedly signifies the vision and goals of APUS, to keep reaching toward the stars and to empower all who use its telescopes.

Bringing Space to the Classroom Through STEM Education Providing Extreme Low Earth Orbit Missions Using ThinSats

Brenda Dingwall and Joyce Winterton
NASA GSFC Wallops Flight Facility

Dale Nash, Sean Mulligan and Brian Crane
Virginia Commercial Space Flight Authority

Robert Twiggs and Matt Craft
Twiggs Space Lab, LLC

Hank Voss and Matt Orvis
NearSpace Launch, Inc.

Abstract

The future of Space Science depends on our ability to attract and engage students into science, technology, engineering and mathematics (STEM) fields. Authentic, hands-on experience with space applications enhances engagement and learning in the STEM disciplines and can help to attract disinterested students to STEM careers. The Virginia Commercial Space Flight Authority (Virginia Space), Twiggs Space Lab, LLC (TSL), Orbital ATK, NearSpace Launch, Inc. (NSL), and National Aeronautics and Space Administration (NASA) Wallops Flight Facility have collaboratively developed the ThinSat Program, providing student teams with the opportunity to design, develop, test, and monitor their own experimental payload that will be integrated into a pico-satellite and launched from the second stage of Orbital ATK's Antares Rocket.

The goal of the program is to provide students with the opportunity to lead and participate in the development of a spacecraft payload through its life cycle over the course of an academic year. The student experience will be enhanced with classroom visits and videos created by the team to educate the students on satellite manufacturing, environmental testing, satellite integration, spaceport, launch vehicle, range, and spacecraft operations. The ThinSat Program will provide a unique and important STEM opportunity for students to develop critical skills in systems engineering and space science that will complement existing programs.

Keywords: STEM education, STEM classrooms, extreme low-orbit missions, Thinsats

Brindando espacio al aula a través de la educación STEM proporcionando misiones de órbita terrestre extremadamente baja usando ThinSats

Resumen

El futuro de la ciencia espacial depende de nuestra capacidad para atraer e involucrar a los estudiantes en los campos de Ciencia, Tecnología, Ingeniería y Matemáticas (STEM). La experiencia auténtica y práctica con aplicaciones espaciales mejora la participación y el aprendizaje en las disciplinas STEM y puede ayudar a atraer estudiantes desinteresados a las carreras STEM. La Autoridad de Vuelo Espacial Comercial de Virginia (Virginia Space), Twiggs Space Lab, LLC (TSL), Orbital ATK, NearSpace Launch, Inc. (NSL) y NASA Wallops Flight Facility, han desarrollado en colaboración el Programa ThinSat, brindando a los equipos de estudiantes la oportunidad para diseñar, desarrollar, probar y controlar su propia carga útil experimental que se integrará en un pico satélite y se lanzará desde la segunda etapa del cohete Antares de Orbital ATK.

Palabras clave: Educación STEM, aulas STEM, misiones extremas de órbita baja, Thinsats

通过STEM教育将太空带到课堂 用ThinSat计划提供极度近地轨道任务

摘要

“空间科学”的未来取决于我们在吸引学生和让其参与科学、技术、工程与数学（STEM）领域的能力。在空间应用上的实际动手经验能提高STEM学科中的参与和学习，并能帮助吸引没有兴趣的学生开启STEM事业。弗吉尼亚州商业航天局（Virginia Space）、Twiggs太空实验室，LLC （TSL）、Orbital ATK、NearSpace Launch公司（NSL）、和国家航空航天局（NASA）沃勒普斯飞行设施，共同协作开发了ThinSat计划，为学生团体提供机会设计、开发、测试和监督其各自的实验装备，后者将被整合到一个皮卫星中，并于Orbital ATK Antares火箭的第二阶段发射。

关键词：STEM教育，STEM课堂，极度近地轨道任务，Thin-sats

Introduction

Reaching students earlier in their educational development cycle is critical in the development of a workforce for the United States so that it can remain competitive in the global marketplace. Teachers in K-12 education must engage students in science, technology, engineering and mathematics (STEM) curriculum earlier to generate interest, develop skills, and provide an educational foundation upon which students can build. The ThinSat Program provides students in middle school, high school, and university with the opportunity to learn many valuable STEM skills that can be applied to future learning opportunities and workforce development. This is critical since the aerospace industry is facing a wave of retirements, with 18.5% of the workforce eligible for retirement in 2017 (Zillman 2013). The shortage of scientists and engineers will significantly impact the aerospace and defense industries' ability to deliver critical technologies necessary to maintain our technological, economic, and military leadership throughout the world.

Companies within the aerospace and defense industries are attempting to address these issues by establishing closer relationships with programs known for producing STEM talent. Outreach programs to universities and establishing research centers are two solutions to the problem. Another solution was the development of the CubeSat Program, which emerged from the university community in 1999. "[T]he CubeSat Program was conceptualized as a tool to help teach students about the process involved in developing, launching, and operating a spacecraft. In 1999, California Polytechnic State University's Multidisciplinary Space Technology Laboratory and Stanford's Space Systems Development Laboratory began discussing ways to provide students with hands-on experience in the field of spacecraft design" (Toorian, Diaz, and Lee 2008). Anecdotal information from universities in the United States and around the world "indicates that hands-on, project-based education is very effective for recruiting, retaining and training engineering students" (Jayaram and Swartwout 2010). The importance of programs like the CubeSat Program is reflected in national priorities identified by federal agencies.

For years, academia has been unable to fully participate in the space revolution due to high costs, launch availability, and a commitment to long-term satellite missions. The invention of the CubeSat offered an initial solution to these challenges and sparked many successful CubeSat programs. The success of small satellites in the space sector has been driven in large part by technological innovations developed through CubeSat programs around the world. As of January 8, 2017, a total of 580 nanosats have been launched. All of this success has created new challenges for university-based CubeSat programs. Many of these universities are transitioning from an educational program to research and development programs and are developing CubeSats that carry instruments and new technologies for other organizations. As a result, universities are busy writing more proposals to secure funding to ensure continuity in their programs. Additionally, the increasing complexity of CubeSats has increased development time from idea to functioning satellite to about three or four years. The problem is that students may not be present for the full development cycle and may miss critical aspects of the process. It is important for students to be involved in the development of spacecraft from conception to operations and to go through all the phases of systems engineering, including mission definition and requirements, system and subsystem requirements and components, designing these systems, and testing and verification against requirements. The CubeSat Program has been instrumental in igniting a passion for students' interest in STEM. The next breakthrough and step forward in engaging students of all ages and fields in STEM programming is the ThinSat Program. It is a cost effective, short-term program that provides students with an exciting opportunity to conduct valuable scientific space-based research.

The goals of the ThinSat Program are to address many of the challenges created due to the success of the CubeSat Program and provide new opportunities for students, including (1) decreasing the spacecraft development cycle time, (2) reducing the complexity and increasing reliability, (3) providing regular launch opportunities, thereby increasing space access, (4) engaging students earlier in their education (fourth to twelfth grade), (5) reducing the burden of paperwork and licensing requirements, (6) mitigating the threat of space debris with short orbital life, (7) reducing the overall cost of spacecraft development and access to space, (8) creating a precursor program to CubeSat programs, and (9) creating a smaller spacecraft platform for valuable space research.

Virginia Commercial Space Flight Authority (Virginia Space), in partnership with Twiggs Space Lab, LLC (TSL), Orbital ATK, NearSpace Launch, Inc. (NSL), and the National Aeronautics and Space Administration (NASA) at Wallops Flight Facility, has developed an educational outreach initiative presented as the ThinSat Program. The purpose of this program is to teach students in middle school and later grades about the iterative engineering design process, systems engineering, data collection methods and analytical processes, and atmospheric and space science. This is achieved by building FlatSats (see Figure 1), performing balloon flight operations, designing and building ThinSat payloads to be launched into orbit, and reviewing satellite data for analysis and reporting to the educational and scientific communities. The inaugural flight, as a secondary payload aboard the second stage of the Orbital ATK Antares rocket, is planned for a mission in the fall of 2018. Subsequent flights will occur every six to twelve months. The secondary payload consists of a 12U volume with the capacity to deploy a maximum of 84 ThinSats at an orbit of approximately 200 to 250 kilometers, allowing for a predicted orbital life of five days. The altitude range of 100 to 300 kilometers is formally referred to as extreme low earth orbit (ELEO), a section of the atmosphere that has not received much attention due to limited orbital lifetimes. This presents an excellent opportunity for students, who are constrained by the academic schedule, to benefit from a short duration satellite program. Additionally, a nominal Antares launch cadence of two or more per year will provide students with the advantage of potentially participating in the ThinSat Program multiple times throughout their academic career.

Prior to the launch of ThinSats as a secondary payload, there will be a structured program to guide teachers and students through the process of developing and testing satellite hardware and data review. The structure contains three phases: *Phase 1*: Sensor development and low altitude balloon flights; *Phase 2*: Payload development, high altitude balloon flights and data review; and *Phase 3*: Flight testing, integration, launch preparations, and final reports. These phases will ensure that students learn the fundamentals of science, engineering, and technology necessary to test and launch a satellite into orbit. This is an excellent opportunity to expose students to the space industry allowing them to learn about engineering processes. The ThinSat Program will inspire the next generation of space scientists and engineers.

ThinSat Program: Three phases

The ThinSat Program is comprised of three phases that will start fourteen months prior to the relative Antares launch date. The three-phase program culminates with participating schools building a ThinSat payload to be delivered to ELEO. Students have two options when developing their payloads. The first payload option utilizes a kit of provided sensors that are common to all ThinSats. A second option is for an institution to develop and create a unique, user-defined payload that meets all ThinSat requirements. The phase breakdown is as follows: Phase 1 of the program will introduce the concept of designing a satellite with easy 'plug and play' electronic sensor chips and launch of low altitude balloons to gather data up to 10 kilometers; in Phase 2, students will use data obtained from the initial flights to design a payload representative of the flight model to be integrated into a ThinSat engineering model (EM). The EM will test the operation and house the payload for a high-altitude balloon flight, which go up to an altitude of 36 kilometers; in Phase 3, data will be analyzed from the prior two balloon flights with the purpose of developing and finalizing a flight payload that will be launched on the Antares. Data from each phase will be stored and displayed real-time through a Space Data Dashboard Interface. Students will access this platform on the Virginia Space website (www.vaspace.org). A compilation of graphs and information will form a personalized dashboard interface for participating institutions. The Space Data Dashboard will also allow schools to collaborate, share information, and upload presentations and reports.

Phase 1A

In Phase 1A of the program, students are introduced to sensors, software, electronics, and data collection methods. The students design, construct and test various configurations of a FlatSat. Each participating school will receive ten electronic satellite kits, which include multiple sensors, power supplies, and related software, manufactured by XinaBox, Inc. (XinaBox). The chips have a user-friendly "plug and play" connection method to form a FlatSat. The small square connectors between the X-chips make it very easy to connect and no soldering is required. X-chips are preprogrammed to ensure students who are not familiar with coding still have the ability to build a FlatSat. Those students who are interested in coding have the option to program the X-chips. In both scenarios, students are exposed to the software and the effects of programming in real-world applications. Once programmed, the X-chips will be tested and operationally verified by using the Wi-Fi or USB modules to connect to an online dashboard, where information will be displayed real-time and stored.

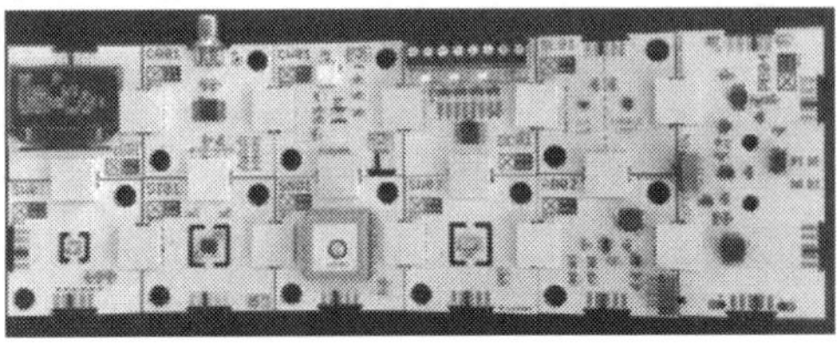

Figure 1: FlatSat/X-Chips

The XinaBox kits include:

1. **Sensors to build ten FlatSats**
    a. USB programmer
    b. Programmable Wi-Fi unit
    c. OLED display chip for sensor data
    d. Light sensor including UVA and UVB
    e. Weather sensor (temperature, humidity, and altitude)
2. **Two balloon mission packs with:**
    a. Long-range transceiver (915 MHz) and monopole antennae
    b. Burn wire module (burn wire unit to run four stages of mission)
    c. Battery pack power supply unit (2XAA)

d. GPS sensor

e. Inertial motion unit (accelerator, magnetometer, and gyroscope)

f. Light sensor (UVA and UVB)

g. Weather sensor (temperature, humidity, and altitude)

h. Large party balloons and parachutes

3. **One ground station to collect data from balloon missions.**

   a. Long-range transceiver (915 MHz) and monopole antenna

   b. USB data connectivity

   c. Programmable Wi-Fi unit

Phase 1B

The FlatSat developed during Phase 1A will be used for the low altitude balloon flights in Phase 1B. As shown above, each school will be provided two balloon mission packs and one ground station to collect data from the FlatSat payload during the low altitude balloon flight. The purpose of this flight is to create opportunities for teamwork among students through mission planning, assignment of roles and responsibilities, and data analysis. Collaboration between schools is encouraged and made possible with the dashboard. The dashboard interface allows students to share and compare results with every institution involved in the mission and between missions.

In addition to the low altitude balloon flight, this phase offers an opportunity to fly FlatSats on a Unmanned Aerial Vehicle (UAV). The Virginia Space UAV Airfield offers great advantages to study the atmosphere at different altitudes and weather conditions, utilizing multiple Unmanned Aircraft System (UAS) platforms.

Figure 2: Virginia Space UAS Airfield

Key concepts and principles to be learned during this phase:

1. **Buoyancy**
2. **Drag and effect of parachutes**
3. **Ideal gas law**
4. **Radio signals**
5. **Temperature and humidity relative to altitude**
6. **Data analysis**

Phase 2

With analysis of the initial flights in Phase 1B, students will design and test a payload utilizing the ThinSat EM. Students can select from motherboard and space-hardened X-chips, which are manufactured with material to withstand the harsh space environment, or they can develop a customized payload from scratch. This will provide students with an initial understanding of how the payload will ultimately have to be designed for the flight model. Once the payload is tested, the students will integrate the payload with the EM and prepare it for a high-altitude balloon fight. This flight will be performed by NSL in Upland, Indiana. NSL will provide four launches with tracking, recovery, and live camera video availability for up to thirty EMs per launch. Students will receive data real-time from each EM utilizing the GlobalStar Network.

ThinSat Engineering Model (EM)

The students build an EM in Phase 2 of the ThinSat Program. This functional EM of the ThinSat is to be used as a teaching model for the ThinSat Flight Model (FM), the interface between balloon launch and ground station, and for testing the direct connection between the flight processor software and hardware. Mechanically, the ThinSat printed circuit board (PCB) will be form-fit-function to the Generation 1 ThinSat FM unit so that all student experiments can be checked with the actual flight processor, electrical interface, ThinSat software, and Space Data Dashboard. No battery, solar arrays, or communication unit will be included. A USB connector is used for ThinSat power and a diagnostic port is included.

A three-axis accelerometer, a three-axis magnetometer, and three-axis gyros (IMU) are included for student testing and learning. Student analog and digital Input / Output (IO) are available (just like the FM). Mechanically, the main structure size and mounting screws will be available as on the FM and will be suitable for the balloon vibration and thermal vacuum environment. A full 3-D printed frame is included with flight shape, viewports, and mounting holes. This frame will be made with durable white nylon and selective laser sintering. A polycarbonate clear cover plate will be used on the –Z plate side instead of the solar array PCB. The student payload area will be the same as the FM so that flight model student experiments can be tested first in the EM as a simulator for the FM. The EM will be delivered to TSL and tested to make sure they are ready for seamless balloon and FM integration.

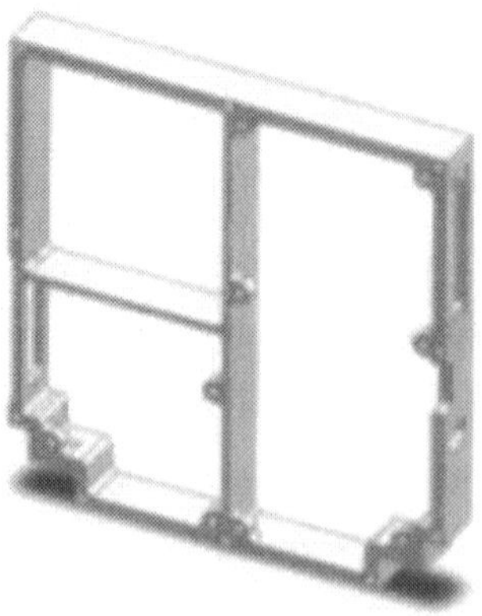

Figure 3: 3-D printed satellite frame

Figure 4: Frame with X-chip motherboard

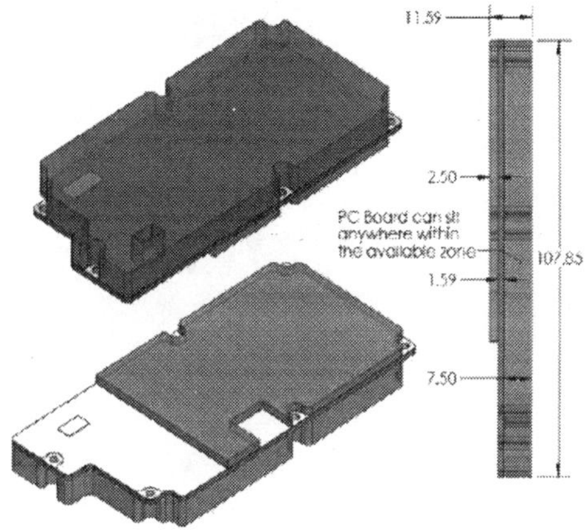

Figure 5: Mechanical drawing of the student payload

Phase 3

After completion of the high-altitude balloon flight, student payloads will be sent back to lead institutions. Students will review flight data and determine if the performance of the payload requires modification. Once the payload is selected, students will send their payload to TSL for environmental testing. The payloads are then integrated into the ThinSat FM and loaded into the Containerized Satellite Dispensers (CSDs) developed by Planetary Systems Corporation.

CSDs loaded with the ThinSats are sent to Orbital ATK for integration onto the second stage of the Antares prior to launch. The ThinSats will then be deployed into ELEO after Second Stage/Cygnus separation. Once in contact with the sun's light waves, the charged batteries will activate and start sending data to the Space Data Dashboard through the Globalstar Network. Students will monitor the data as the ThinSats slowly deorbit and eventually burn up. After completion of the ThinSat mission, students will perform a final data analysis and report their results.

ThinSat as a Picosatellite

The ThinSat is a picosatellite that weights approximately 280g and has dimensions of 111 x 114 x 17.4 millimeters. Of these dimensions, approximately 50 percent of the volume will be reserved for student payloads. This area allows space for a customized payload or six X-chip sensors and a motherboard. The ThinSat dimensions are based on the CubeSat form factor. A volume of 1U or a ten-centimeter cube is equal to seven ThinSats. Therefore, in each 3U CSD, there are twenty-one ThinSats. Among the twenty-one ThinSats there will be groups tethered together to form multiple strings, based on multiples of three (3, 6, 9, etc.). The number of strings may differ in separate CSDs; for example, one 3U CSD may have three strings of six and one string of three ThinSats, while another CSD may be comprised of seven strings of three ThinSats. The number of strings determined will be based on relative mission requirements and payload specificity. Each string will have one mothership and subsequent daughter-ship(s). See the ThinSat ICD (Dailey, Orvis, and Voss 2017) for more technical information on the ThinSat payload and operation.

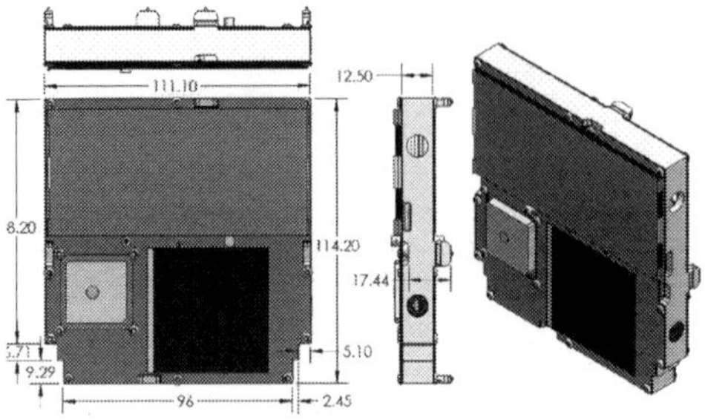

Figure 6: ThinSat Dimensions in Millimeters

Figure 7: ThinSat Configuration for Deployment

Mothership Hardware

The mothership ThinSat of a string will contain the following hardware:

1. **Student Payload**
   a. Customized payload
   b. Space-hardened X-chips
   c. Sensors provided manufactured by NSL
2. **Globalstar Radio**
3. **Flight Processor**
4. **Piksi GPS**
5. **Foldout Camera**

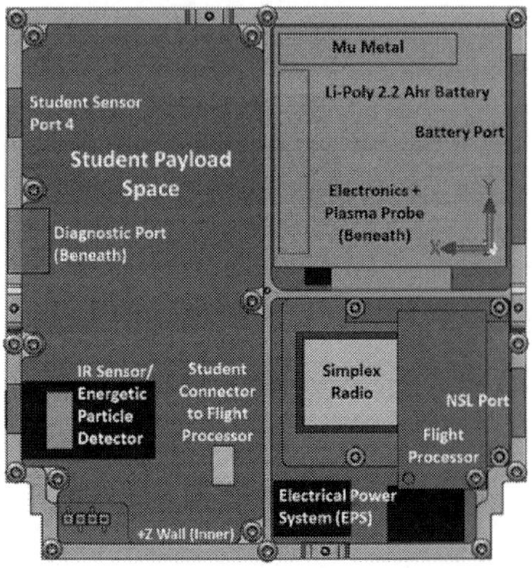

Figure 8: Mothership Space Utilization

Daughtership Hardware

The first daughtership will supplement battery power to the mothership, which demands more battery power. The daughterships will contain the following:

1. **Student Payload**
   a. Customized payload
   b. Space-hardened X-chips
   c. Sensors provided, manufactured by NSL
2. **Globalstar Radio**
3. **Energetic Particle Detector**
4. **Plasma Probe Board**
5. **Flight Processor**

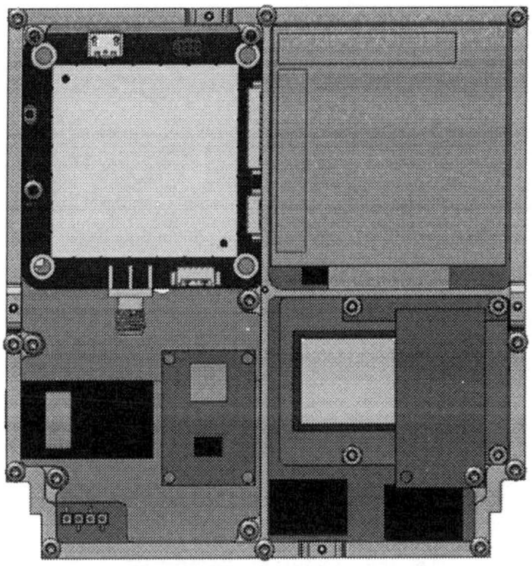

Figure 9: Daughtership Space Utilization

ThinSat Grouping Types and Tethering Options

The ThinSats will be grouped according to mission requirements and payload specificity. The tethers used to group the ThinSats will be either nitinol wire or nitinol ribbon with solar arrays attached.

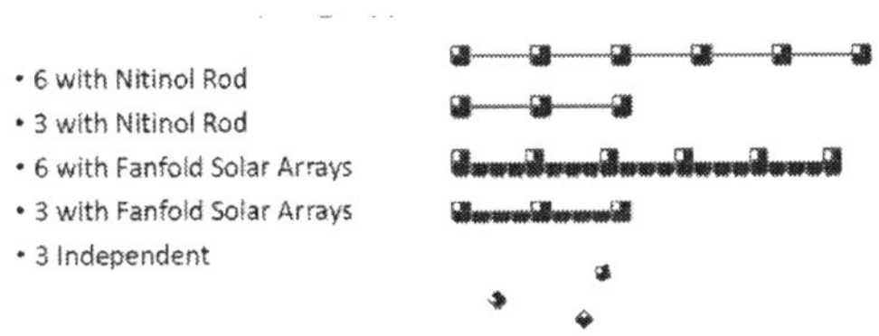

Figure 10: ThinSat Grouping Types

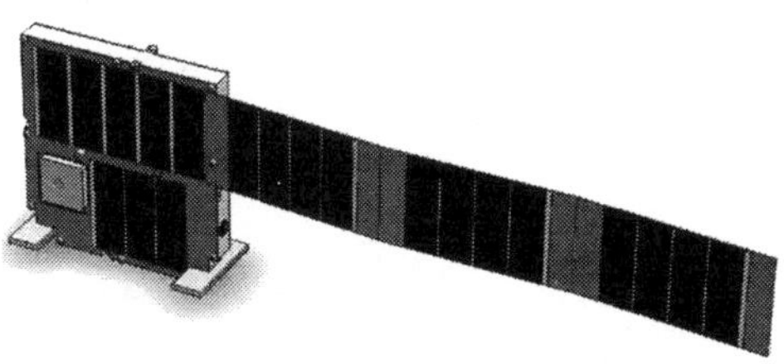

Figure 11: ThinSat Fanfold Solar Array

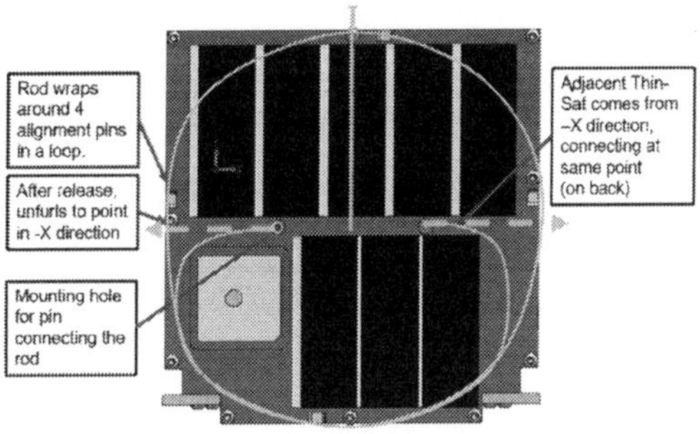

Figure 12: Nitinol Wire Configuration

Space Data Dashboard

The ThinSat Program will have a user-friendly Space Data Dashboard that will act as a focal point for data analysis and collaboration among participating schools. This interface will be accessible through the Virginia Space website and will allow institutions to create plots for sensors being used during the program. The plots will be populated during sensor testing, balloon flights, and orbital flight. Data will be plotted real-time and stored for accessibility at any time. Student data from each mission will be stored and open for the entire ThinSat community, allowing students to view, compare, and analyze datasets from any institution's flight.

Aside from viewing and storing data, the Space Data Dashboard will act as a "one place" reference for all the information on the ThinSat Program. There will be a wiki sub-page for information regarding the program phase instructions for teachers and students, past analysis reports from participating schools, and a collaboration board or blog where questions or results can be posted.

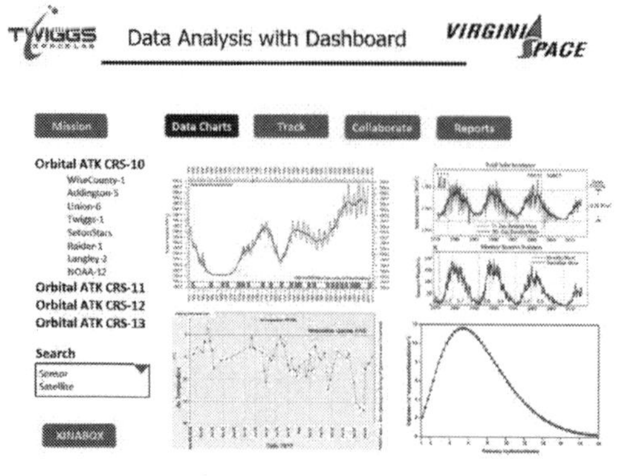

Figure 13: Space Data Dashboard Interface

Education and Curriculum

The curriculum developed for the ThinSat Program is intended for students in middle and high school. The same guidelines and principles are followed for each of the three phases, but a broad range of STEM subjects and sophistication levels can be studied depending on the complexity of the student payload. For the standard X-chip sensors, a user-friendly model is prepared so younger students may participate in the program. Students in eleventh grade through college may want to develop a separate payload; therefore, a structured curriculum need not be followed. Instead, the students can prepare a midterm and final report on the reason they chose the specified payload, the data accumulated, and the purpose of the findings.

The ThinSat Program will use a club-style approach, where each school will determine the number of meetings. A recommended curriculum was developed using the Next Generation Science Standards (NGSS) and State Developed Standards to incorporate required topics into the ThinSat Program. An assessment test will be provided at the start and end of the program to track student progress and ensure student learning.

The Science Standards Topics to be Covered by the ThinSat Program Curriculum

Earth Space Science

1. **Weather and Climate**

   Students will collect and analyze data from multiple locations, across the nation and globe. This allows students a unique opportunity to examine different weather systems and how to explain atmospheric data relative to location and time.

2. **Space Systems**

   In phase 3, students will explore the effects of microgravity, orbital life, orbit inclination, and solar flux on the ThinSats. This will allow for an understanding of why the ThinSat deorbits at different time intervals, but always close to a week.

3. **Earth Systems**

   To incorporate earth systems, students will investigate the water cycle with the temperature, humidity and pressure using X-chips. This data will be analyzed relative to altitude and atmospheric density.

Life Science

1. **Matter and Energy in Organisms and Ecosystems**

   Through data analysis, students will be able to gain an understanding of Earth's biotic and abiotic factors. They will also be able to understand and explain human impacts on various ecosystems.

Engineering, Technology, and Application of Science

1. **Engineering Design**

   The ThinSat Program as a whole represents the iterative engineering process. Before the final ThinSat payload is manufactured, students use Phase 1 to understand how the sensors work through various tests and a low altitude balloon flight. Next, they use an EM to design and test a payload using the High-Altitude Balloon Flights. If everything is nominal and no modification is required, the students have completed building and testing a successful payload that is ready for space flight testing and ultimately orbiting Earth.

Phase:	Duration:	# of Sessions:	Classroom/Club Hours:
1	15 weeks	15-30	15-45
2	14 weeks	14-28	14-42
3	8 weeks	08-16	08-24

Table 1: Curriculum Schedule Breakdown

Curriculum structure within ThinSat Program phases to incorporate Science Standards

To incorporate the standards above, each phase will be broken down into sections, which last anywhere from two to five weeks. It is anticipated that schools in the program will learn at different paces; thus, flexibility is provided as long as the phase is completed on time and all section requirements have been satisfied.

Phase 1: Sensor Development and Low Altitude Balloon Flights

1. **Essential Questions:** Use NGSS and State Developed Standards to generate interest in topics
2. **Research Window:** Determine methods to answer essential questions with hardware provided
3. **Design and Manufacture:** Build FlatSat payloads to obtain data to help explain essential questions
4. **Testing Protocol:** Test FlatSats to ensure usability
5. **Final Product:** Use final FlatSat payload on low altitude balloon flights

Phase 2: High Altitude Balloon Flights and Data Analysis

1. **Essential Questions:** Develop questions in topics related to the high-altitude balloon flight
2. **Research Window:** Determine methods needed to answer essential questions
3. **Design and Manufacture:** Build payload with dimensions of ThinSat flight payload
4. **Testing Protocol:** Test payload using ThinSat EM
5. **Final Product:** Use payload and EM on high-altitude balloon flight

Phase 3: Flight Testing, Integration, Launch Preparations, and Final Reports

1. **Essential Questions:** Develop

questions in topics related to launch and flight of the ThinSats in orbit.

2. **Testing Protocol:** If necessary, make modifications to payload from Phase 2. Test payloads for flight in orbit.
3. **Data Analysis and Final Reports:** Analyze data from the ThinSats in orbit and develop a final report.

Timeline

For any selected launch, there will be a group of lead institutions, universities, or professional institutions that enlist local schools. It is the responsibility of lead institutions to assist and guide schools they involve in the program, through the fourteen-month, three-phase program. Designating the Antares launch date as "L," the timeline of events proceeds as follows:

L-14 Months: Delivery of XinaBox Kits

L-14 Months: Start of Phase 1

Students will use the XinaBox kits delivered at L-14 to program and design FlatSats. Low-altitude balloon and UAV flights to be conducted, data analyzed, and payloads modified.

L-10 Months: Start of Phase 2

Modified FlatSat payloads are designed to fit in the student payload space of the ThinSat EM. The model will be used to test payload and verify data transmission. The payloads are then integrated and secured within the EM in preparation for the high-altitude balloon flight. Students then send the EM with the payload integrated to NSL for the high-altitude balloon flight.

L-06 Months: Start of Phase 3

After completion of the high-altitude balloon flight, the EMs with student payloads are sent back to the lead institutions, who analyze data to make changes to the payload if necessary. Final modifications are made to the student payloads and sent to TSL for orbital flight testing.

L-04 Months: TSL Testing of ThinSat Flight Models

TSL conduct the environmental testing of the spacecraft at the Morehead State University's Spacecraft Environmental Testing Laboratory (SETL), located within the Space Science Center, which provides for testing and qualification services for spacecraft up to 100 kg. The SETL is capable of supporting Hardware in the Loop (HWIL) testing to NASA GEVS level and greater. The SETL has a rich heritage of testing and qualifying in-house built satellites and is available as a commercial service for both public and private sectors. The space environmental testing includes:

1. **EMI/EMC Testing**

   Complete EMI/EMC Testing to MIL-STD-461C: Electromagnetic emission and susceptible requirements for the control of electromagnetic interference.

2. **Vibration Testing**

   Vibration testing verifies satellite

survivability post-launch and can identify mechanical and structural faults and stresses. The SETL's vibration slip table allows for three axes of testing at or above NASA GEVS levels and can be customized per mission ICD.

3. **Thermal Vacuum (T-Vac) Testing**

   Thermal vacuum (T-Vac) testing verifies satellite performance in a simulated space environment with temperature extremes beyond what the satellite is expected to experience on orbit. The SETL's T-Vac system has a capacity of 0.29 m^3 (10 ft^3) and a temperature range of -100°C to +220°C at 1x10-8 torr. Pass-throughs allow for functional testing under vacuum.

L-01 Month: CSDs with ThinSats are sent to Orbital ATK for integration with Antares Second Stage

L-00: Launch of the Antares

L+10 minutes: Second Stage-Cygnus Separation

L+12 minutes: Deployment of ThinSats

L+05 days: Estimated ThinSat orbit lifespan

The orbital lifespan of the ThinSats varies depending on atmospheric parameters.

L+1-3 Months: Final Report Due

Conclusion

The ThinSat Program was specifically developed to engage students of all ages in STEM. In order to follow academic schedule guidelines and reach a broad profile of students, this program provides a flexible curriculum. Teachers will not be constrained by single lesson plans, but rather will be provided curriculum guides that introduce general concepts and how they can be studied. The ThinSat Program provides an approachable and recurring opportunity each academic year for students and teachers to collaboratively participate in hands-on space science and engineering, opportunities that were previously reserved for research universities with long time horizons. This exposure will hopefully encourage students to enter into STEM fields and ultimately help foster the next generation of scientists and engineers in the aerospace arena. Their participation in the ThinSat Program could provide both the spark of engagement and practical application of the scientific method to help prepare the next generation. ThinSats will allow students of all grades to get their fingerprints into space, providing a positive impact on the vitality of the industry and the future of space exploration.

Acknowledgments

We would like to thank the Virginia Space Consortium for the support and funding of this project.

References

Dailey, J., M. Orvis, and H. Voss. 2017. "ET-Sat Student Payload Interface Control Document." www.NearSpaceLaunch.com.

Jayaram, S. and M.A. Swartwout. 2010. "A Review of the Role of Student-Built Spacecraft in Workforce Training and Innovation: Ten Years Of Significant Change." In *AIAA Space 2010 Conference & Exposition.* https://doi.org/10.2514/6.2010-8735

Toorian, A., K. Diaz, and S. Lee. 2008. "The CubeSat Approach to Space Access." *Aerospace Conference, 2008 IEEE.* https://doi.org/10.1109/AERO.2008.4526293.

Zillman, C. 2013. "The Average Age of Aerospace Engineers in the U.S. is 47, and Many of these Jobs can't be Filled by Foreign Workers." *Fortune.* http://fortune.com/2013/11/12/americas-defense-industry-is-going-gray/.

Modeling and Simulation of a Long-Wave Infrared Polarimetric Sensor for Space Object Detection and Characterization

Kevin Pohl, Jonathan Black, Jonathan Pitt and Edward Colbert
Virginia Polytechnic Institute and State University

Abstract

Long-wave infrared (LWIR, wavelength > 8 um) polarimetric measurements can be used to characterize space objects under certain conditions. Both visible and LWIR polarimetry have been demonstrated extensively in terrestrial applications for characterization and detection of objects of interest. Visible polarimetry has also been demonstrated for space object detection. A simulation of a camera and telescope for collection of LWIR polarimetric signatures of space objects has been assembled using three software packages: Systems Tool Kit (STK), MATLAB, and FRED. Characterization of space objects is generally possible across a wide range of target surface temperatures and emissivities, and at a sub-pixel level; characterization is reliable in a narrower range. This approach represents an initial step forward in optical systems for space situational awareness (SSA) in that it offers a wider field of view than equivalently sized visible light collectors, and it can be used both day and night, regardless of target illumination.

Keywords: modeling, simulation, long-wave infrared, polarimetric sensor, space objects, detection, characterization

Modelado y simulación de un sensor polarimétrico infrarrojo de onda larga para la detección y caracterización de objetos espaciales

Resumen

Las mediciones polarimétricas de infrarrojos de onda larga (LWIR, longitud de onda> 8 um) se pueden utilizar para caracterizar objetos espaciales bajo ciertas condiciones. Tanto la polarimetría visible como la LWIR se han demostrado ampliamente en aplicaciones terrestres para la caracterización y detección de objetos de interés.

La polarimetría visible también se ha demostrado para la detección de objetos espaciales. Se ha ensamblado una simulación de una cámara y un telescopio para la recolección de firmas polarimétricas LWIR de objetos espaciales utilizando tres paquetes de software: Systems Tool Kit (STK), MATLAB y FRED. La caracterización de los objetos espaciales generalmente es posible en una amplia gama de temperaturas y emisividades de la superficie objetivo, y en un nivel subpíxel; La caracterización es confiable en un rango más estrecho. Este enfoque representa un paso inicial en los sistemas ópticos para la conciencia de la situación espacial, ya que ofrece un campo de visión más amplio que los colectores de luz visible de tamaño equivalente, y se puede usar tanto de día como de noche, independientemente de la iluminación del objetivo.

Palabras clave: modelado, simulación, infrarrojo de onda larga, sensor polarimétrico, objetos espaciales, detección, caracterización

一个用于空间物体探测和定性的长波红外线偏振式传感器的建模和模拟

摘要

长波红外线（LWIR，波长大于8微米）的偏振测量能在一定条件下用于描述空间物体的特征。可见和长波红外线偏振测量已在地面应用中被广泛用于描述和探测研究物体。可见偏振测量还被用于探测空间物体。通过使用三种软件包：Systems Tool Kit（STK）、MATLAB和FRED，组装了一个模拟相机和望远镜，用于收集空间物体的长波红外线偏振特性。以一个大范围的目标表面温度和发射率来描述空间物体的特征通常是可能的，并且从亚像素层面也是可能的；在更窄范围内进行特征描述是可信赖的。此方法代表了就空间场景意识而言光学系统中的初步成果，比起相同尺寸的可见光收集器，此方法提供了一个更广的视角领域，同时在白天和夜间都能使用，与目标照射无关。

关键词：建模，模拟，长波红外线，偏振式传感器，空间物体，探测，特征化

Introduction

The objective of the project, the opening stage of which is documented here, is to explore a new avenue for space situational awareness (SSA). Today, SSA is maintained through a combination of ground sensors (radars, telescopes) and voluntary sharing of telemetry and other information with various organizations that maintain active catalogs of space objects. The present work seeks to demonstrate, initially through modeling and simulation, the characterization of low Earth orbit (LEO) satellites by an LWIR polarimetric imaging system.

LWIR polarimetry has been demonstrated for man-made object detection in a variety of settings (Gurton et al. 2010; LeMaster & Eismann 2014). In addition, visible spectrum polarimetry has been demonstrated for detection and identification of satellites in geosynchronous orbit (GEO) (Speicher 2015; Speicher et al. 2014). Speicher used visible light polarimetry to detect and identify GEO satellites. The experimental setup only measured S0 and S1, and due to the dimness of the targets required an integration time of ~20 seconds. Repeated observations over time revealed differences in signatures between individual satellites, both between different types of vehicles, and between vehicles of the same design, but of different age. The latter effect is of particular interest as it is the material properties of the surface layer (e.g., paint) that drive the complex index of refraction and thus the polarization signature. Those material properties change over time due to exposure to the space environment (Speicher 2015). Further work has shown that the polarization signatures of individual components (e.g. dish antenna, bus, solar panels) sum together to create a composite signature (Beamer et al. 2017, 2018).

Passive LWIR polarimetry for man-made object detection has been the subject of numerous studies since at least the 1990s (Gurton et al. 2010; LeMaster & Eismann 2014; Rogne et al. 1990) The advantage of LWIR for these purposes is that it measures primarily the target's self-emission, though emission by nearby sources (e.g., low clouds), and thus reflection off the target, can interfere. This occurs because the reflected light is polarized perpendicularly to the emitted light, resulting in reduced values for S1 and S2 (Rogne et al. 1990; Wellems et al. 2006) and thus a reduced signal-to-noise ratio. For space object detection, this is not a concern, as space objects are generally not close to each other, and solar radiation on the target (and resulting reflected radiation) in the 8-9 micron band is an order of magnitude or more less than the self-emission of the target.

Ground target detection using LWIR polarimetry is effective both day and night. An object viewed from a specific angle will remain virtually unchanged in terms of degree of linear polarization (DOLP), regardless of changes in thermal contrast with its surroundings (Gurton et al. 2010)

There has also been some work done concerning LWIR (without polarization) for detection of space objects, beginning in the late 1980s. Targets included geosynchronous satellites, which could not be resolved, but could be detected (Baddiley 1990; Lee & Nishimoto 1993; Seniw 1993). Studies have also been conducted using space based LWIR sensors to detect and characterize space objects (McCall et al. 2014). In both cases, detection was feasible during both daytime and nighttime. The tradeoff between visible light collection and LWIR is one of resolution against collection opportunities. When using a long-wave IR sensor, the target is its own source of illumination, where a visible light sensor requires an external source of illumination (e.g., the Sun).

Polarization measurement

Several methods exist to measure the polarization of light, Jones matrices (Fowles 1989), coherency matrices (Azzam & Bashara 1977), Mueller matrices, and others (Chipman 1995). Of these methods, Mueller calculus is most appropriate for use with polarimeters, where the polarization state of a light beam is described by the Stokes vector **S** and the polarization-altering characteristics of a target are described by the Mueller matrix **M**. The Stokes vector is based on six flux measurements using ideal polarizers in front of a radiometer: horizontal (P_H), vertical (P_V), diagonal (45 and 135 degrees; P_{45} and P_{135}, respectively), and left (P_L) and right circular (P_R) (Chipman 1995).

The Stokes vector is then defined as

$$S = \begin{bmatrix} s_0 \\ s_1 \\ s_2 \\ s_3 \end{bmatrix} = \begin{bmatrix} P_H + P_V \\ P_H - P_V \\ P_{45} - P_{135} \\ P_R - P_L \end{bmatrix} \quad (1)$$

where s_0, s_1, s_2, and s_3 are the Stokes vector components in units of watts per meter squared. The Stokes vector represents an average over area, solid angle, and wavelength (Chipman 1995). From the Stokes vector, four polarization parameters can be determined (Azzam & Bashara 1977):

Flux $$P = s_0 \quad (2)$$

Degree of polarization $$DOP = \frac{\sqrt{s_1^2 + s_2^2 + s_3^2}}{s_0} \quad (3)$$

Degree of linear polarization $$DOLP = \frac{\sqrt{s_1^2 + s_2^2}}{s_0} \quad (4)$$

Degree of circular polarization $$DOCP = \frac{s_3}{s_0} \quad (5)$$

Of these, flux and DOLP are most relevant for the present study. The bulk of the materials encountered–dielectrics, metals, and thin films (coatings, paints)—have negligible rates of circular polarization (Chipman 1995), reducing the value of DOP and DOCP measurements.

Tools

Systems Tool Kit (STK)

STK provides a sophisticated modeling environment to model space systems and evaluate their performance. For this project, STK provides the ability to model the positions and attitudes of a satellite with respect to a notional ground-based telescope over time: range to target and angle from which the satellite is seen by the telescope. In addition, the EOIR (electro-optical/infrared) toolkit can provide simulated imagery based on the selected parameters for sensor and target, though it is not capable of simulating polarimetric measurements.

Optical Photonics: FRED

FRED is a software suite that simulates the propagation of light through any optomechanical system by raytracing, including polarimetric measurements. FRED keeps track of the polarization of the light as it makes its way from the target to the sensor.

Modeling and Simulation

The range of temperatures and material emissivities for which an object in space could be expected to self-emit substantially more LWIR radiation than it reflects from incoming solar radiation were determined. This is important, because reflected polarization competes with emitted polarization when calculating the Stokes vector components (Rogne et al. 1990).

To calculate the solar flux on an object at a given wavelength, one must first determine the total spectral radiance of the Sun ($b_{T,Sun}$) at a given wavelength:

$$p = 2 \cdot h \cdot \frac{c^2}{\lambda^5} \tag{6}$$

$$b_{T,Sun} = p/(\exp\left(\frac{h \cdot c}{\lambda \cdot k \cdot T}\right) - 1) \tag{7}$$

where h is Planck's constant, c is the speed of light, k is Boltzmann's constant, T is the temperature of the Sun, and λ is the wavelength. Next, the radiance is multiplied by the square of the ratio of the solar radius to the radius of Earth's orbit and Lambert's cosine law is applied:

$$b_{T,Sun@EarthOrbit} = b_{T,Sun} \cdot (2.177 \cdot 10^{-5}) \cdot \pi \tag{8}$$

The calculation for a notional object in LEO is similar; however, the emissivity of the object in the waveband of interest needs to be factored into the calculation.

$$b_{T,Obj} = p/(\exp\left(\frac{h \cdot c}{\lambda \cdot k \cdot T}\right) - 1) \cdot \pi \cdot \varepsilon \tag{9}$$

where ε is the emissivity of the object at a given wavelength λ.

This process is repeated for 200 discrete wavelengths between 8 and 9 microns and the resulting blackbody radiation values are summed across the waveband of interest. A MATLAB script was written to compare self-emission with incoming solar radiation in the 8 to 9 μm band. The resulting graph (Figure 1) shows that for all but the most reflective surface materials (e.g., bare, polished metal), even relatively cold objects (below 270K) emit more than an order of magnitude more thermal radiation than they receive from the Sun.

Since rejecting waste heat is an important function of a satellite's outer structure, emissivities of 0.8 or more in the LWIR can be expected. In addition, active LEO space objects will spend most of their lifetime on the warm side of 270K—active satellites in general can be expected to have surface temperatures between 270K and 380K (Seniw 1993). Because the target's self-emis-

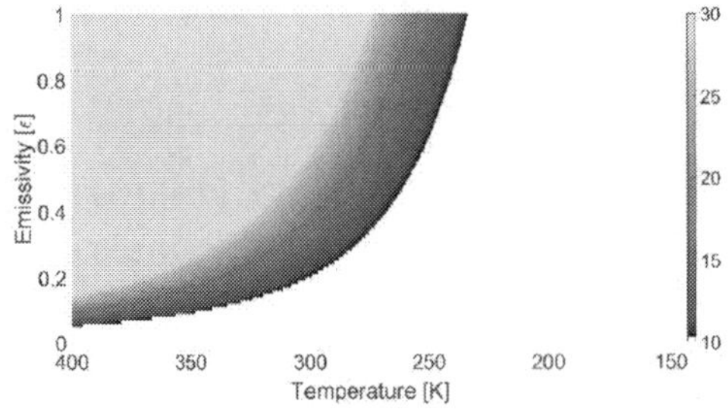

Figure 1: Comparison of self-emission to reflected solar radiation as a function of object emissivity and temperature (the color scale maximum is 30, but the maximum calculated factor was >400)

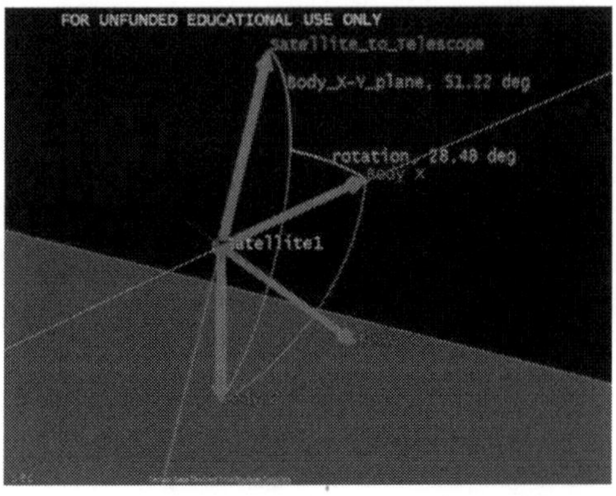

Figure 2: Target body coordinate system (blue arrows), telescope-to-target vector (pink arrow), rotation angles (cyan, green), orbital track (yellow line)

sion is the dominant contribution to this temperature range, this method allows detection whenever the target is overhead, including when the target is in Earth's shadow.

The overall algorithm is a manual workflow using STK to generate range and rotation angle parameters to subsequently use as input in computing polarimetric measurements in FRED. A simplified model the size of the International Space Station (ISS) was placed in an appropriate orbit in STK using publicly available two-line elements, and the orbit was propagated for one month to ensure a wide variety of viewing angles and target distances. Figure 2 shows a view of the orbit track, the target body coordinate system (blue arrows), the vector from the telescope to the target (pink arrow), and the two rotation angles that describe how the target will need to be rotated in FRED to simulate the polarimetric measurement.

STK generates a comma-separated-values file listing all the contacts over the course of a month, along with the magnitude of the distance from the telescope to the target and the rotation angles in 60-second increments. A close overhead pass was selected from this dataset to perform an initial set of simulations in FRED.

The telescope design selected for this simulation was a modified Cassegrain-type design with two mirrors and a refractive corrector near the image plane. Figure 3 shows the telescope model in FRED. The primary mirror is 74.7 cm in diameter with a central obscuration of 30.8 cm and f/3.7. The notional sensor array is an HgCdTe framing focal plane with 640x480 pixels and a pixel pitch of 25 μm. The telescope/sensor combination delivers diffraction limited performance. An ISS-sized model was selected as an easy initial test case, as its large size ensures resolved imagery with the selected telescope design at closest approach.

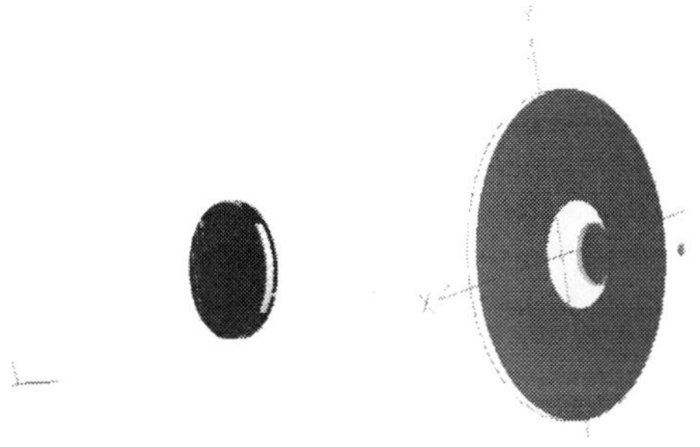

Figure 3: Telescope model in FRED

The target was created as a simple 3-D model in FRED (see Figure 4). Each surface was assigned a material: aluminum for target structural elements, indium gallium arsenide phosphide for the target solar panels, aluminum for the telescope mirror surfaces, and germanium for the telescope lens. Each surface was configured as an emitter in the LWIR band from 8 to 9 μm. Only rays that reach the simulated sensor array are traced.

Figure 4: Target model in FRED

The target model was placed at a distance from the telescope corresponding to the target distance calculated by STK and rotated to correspond to the angle from which the telescope would view the target. For each target configuration FRED then generated an irradiance spread function across the sensor array, effectively a simulated thermal image (Figure 5, Figure 6) and polarization information for each pixel on the array. An embedded script takes that information and determines the aggregate Stokes vector for the target in its current configuration. As expected, the s_3 values were consistently 4 to 5 orders of magnitude weaker than s_0, s_1, and s_2.

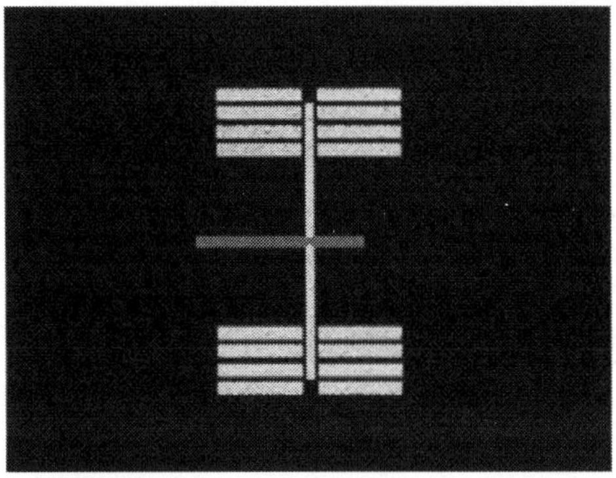

Figure 5: Simulated thermal image of target object

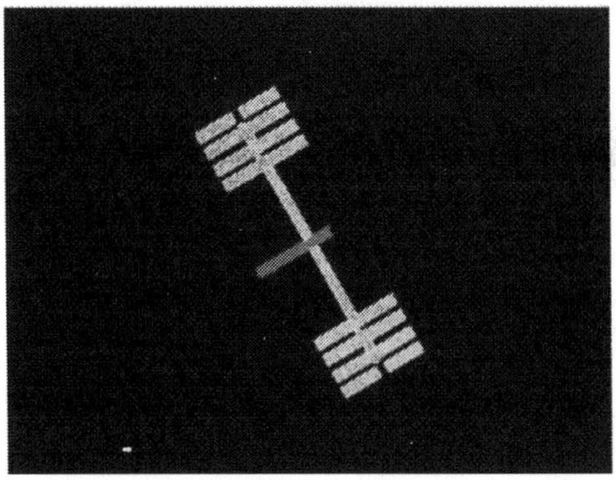

Figure 6: Simulated thermal image of target object rotated about its y- and z-axes (60 degrees and 30 degrees, respectively)

Table 1 shows the results for the two target views presented in Figure 5 and Figure 6. Because the target features a rounded surface for part of its structure, even the straight-on view has some linear polarization. The angled target has a much higher value, however—as expected. Moving the target farther from the sensor reduces the magnitudes of the Stokes vector components, but not their relative values, so that DOLP is unchanged.

	s_0	s_1	s_2	DOLP
Figure 5	0.0210	-2.54×10^{-5}	0.0145	.694
Figure 6	0.0122	0.0103	-0.00596	.979

Table 1: Simulated polarization values for different rotation angles of the target (s_0, s_1, s_2 in W/m^2)

Conclusions

Using two advanced modeling tools, STK and FRED, it has been shown that it is possible to characterize a target at LEO distances and measure its polarization state using a simulated LWIR polarimeter.

The work so far is a proof of concept, showing that a LWIR polarimeter can be used to characterize remote space objects. Key questions for continuing work on the subject are the range of target parameters for which this methodology works; i.e., how small a target, how far away, and to what degree individual objects can be characterized and identified, especially in a scene with multiple targets present.

References

Azzam, R.M.A., & N.M. Bashara. 1977. *Ellipsometry and Polarized Light.* Amsterdam: North-Holland Publishing Company.

Baddiley, C.J. 1990. "The Potential of CdHgTe Staring Array Infrared Detectors for Satellite Detection." In *Proceedings of Optical Systems for Space and Defence,* London, United Kingdom.

Beamer, D.K., U. Abeywickrema, & P.P. Banerjee. 2018. "Statistical Analysis of Polarization Vectors for Target Identification." *Optical Engineering* 57 (5).

Beamer, D.K., U. Abeywickrema, & P.P. Banerjee. 2017. "Statistical Analysis of Polarization Vectors for Target Identification." In *Proceedings of SPIE Optical Engineering + Applications: Polarization Science and Remote Sensing VIII,* San Diego, CA.

Chipman, R.A. 1995. "Polarimetry." In *Handbook of Optics,* Volume II. New York: McGraw-Hill.

Fowles, G.R. 1989. *Introduction to Modern Optics.* New York: Dover Publications.

Gurton, K., M. Felton, M., R. Mack et al. 2010. "MidIR and LWIR Polarimetric Sensor Comparison Study." In *Proceedings of SPIE Defense, Security and Sensing: Polarization: Measurement, Analysis, and Remote Sensing IX,* Orlando, FL, April.

Lee, J.K., & D.L. Nishimoto. 1993. "Infrared Detection of Geosynchronous Objects at AMOS." In *Proceedings of the 1993 Space Surveillance Workshop,* Lexington, MA.

LeMaster, D., & M. Eismann, 2014. "Passive Polarimetric Imaging." In *Multi-Dimensional Imaging,* John Wiley & Sons.

McCall, P.D., M.L. Naudeau, & M. Adjouadi. 2014. "Debris Characterization Techniques Via Unresolved Long-Wave Infrared Imaging from a Space Platform." *Journal of Applied Remote Sensing* 8 (1).

Rogne, T.J., F.G. Smith, & J.E. Rice. 1990. "Passive Target Detection Using Polarized Components of Infrared Signatures." In *Proceedings of Polarimetry '90: Radar, Infrared, Visible, Ultraviolet, and X-Ray,* Huntsville, AL, October.

Seniw, W.P. 1993. "LWIR Observations of Geosynchronous Satellites." In *Proceedings of the 1993 Space Surveillance Workshop,* Lexington, MA.

Speicher, A. 2015. *Identification of Geostationary Satellites Using Polarization Data from Unresolved Images.* University of Denver, Denver.

Speicher, A., M. Matin, R. Tippets et al. 2014. "Calibration of a System to Collect Visible-Light Polarization Data for Classification of Geosynchronous Satellites." In *Proceedings of SPIE Optical Engineering + Applications: Remote Sensing System Engineering V,* San Diego, CA, September.

Wellems, D., S. Ortega, D. Bowers et al. 2006. "Long Wave Infrared Polarimetric Model: Theory, Measurements, and Parameters," *Journal of Optics A: Pure and Applied Optics* 8.

Tailored Systems Engineering Processes for Low-Cost High-Risk Missions

Jared Clements and Tyler Murphy
ATA Aerospace

Lee Jasper
Space Dynamics Laboratory

Charlene Jacka
AFRL/RVEN

ABSTRACT

Given the low cost of most CubeSat missions, a full implementation of the traditional space systems engineering process to CubeSat missions can be detrimental to the programmatic success of the CubeSat. At the other extreme, CubeSat missions often suffer predictable consequences from the omission of standard systems engineering processes, such as risk management, configuration management, and quality assurance. In this paper, we discuss a scaled systems engineering approach to CubeSat missions implemented on a programmatically constrained mission. We also discuss each of the standard systems engineering processes and options for tailoring the processes for a constraint-based mission and how this varies from typical top-down mission processes. The intent is to inform the decisions of mission developers in determining what level of rigor is appropriate for each process in their unique circumstances and mission needs. Examples of tailoring processes utilized with missions currently underway at the Air Force Research Laboratory's Small Satellite Branch (AFRL/RVEN) are used to illustrate the application of the information presented.

Keywords: systems engineering processes, low cost, high risk missions, Air Force Research Laboratory's Small Satellite

Procesos de ingeniería de sistemas a medida para misiones de alto riesgo y bajo costo

Resumen

Dado el bajo costo de la mayoría de las misiones cubesat, una implementación completa del proceso de ingeniería de sistemas espaciales tradicionales para las misiones cubesat puede ser perjudicial para el éxito programático del cubesat. En el otro extremo, las misiones Cubesat a menudo sufren consecuencias predecibles por la omisión de los procesos estándar de ingeniería de sistemas, como la gestión de riesgos, la gestión de la configuración y el aseguramiento de la calidad. En este artículo discutimos un enfoque de ingeniería de sistemas a escala para las misiones Cubesat implementadas en una misión con limitaciones programáticas. Una discusión de cada uno de los procesos y opciones de ingeniería de sistemas estándar para adaptar los procesos para una misión basada en restricciones y cómo esto varía de los procesos típicos de arriba hacia abajo de la misión. La intención es informar las decisiones de los desarrolladores de la misión para determinar qué nivel de rigor es apropiado para cada proceso en sus circunstancias y necesidades de misión únicas. Para ilustrar la aplicación de la información presentada, se utilizan ejemplos de procesos de adaptación utilizados con misiones actualmente en curso en la Subdivisión de Satélites Pequeños del Laboratorio de Investigación de la Fuerza Aérea (AFRL / RVEN).

Palabras clave: procesos de ingeniería de sistemas, misiones de bajo costo y alto riesgo, satélite pequeño del Laboratorio de Investigación de la Fuerza Aérea

为低成本高风险任务定制的系统工程过程

摘要

鉴于大多数立方卫星任务的低成本，对立方卫星任务执行完整的传统空间系统工程过程可能会不利于立方卫星的计划成功。而在另一个极端，立方卫星任务时常因缺少标准系统工程过程（例如风险管理、配置管理和质量保证）而遭受可预见的后果。本文中我们探讨了一个针对立方卫星任务的可扩展系统工程方法，将方法应用于一个受编程限制的任务。探讨了每一个标准系统工程过程，探讨了为一个基于限制的任务定制过程的各选项，探讨了对典型的自上而下任务过程而言如何存在差异。目的是在根据每个过程的独特情况和任务需求确定各个过程的严格程度时，影响任务开发者的决策。当前在空军研究实验室小型卫星部门（AFRL/RVEN）进行的任务定制过程案例被用于阐述上述信息的应用。

关键词：系统工程过程，低成本，高风险任务，空军研究实验室之小型卫星

Constraint-Driven Design

Small satellites see significant utilization because they are intended to be both lower-cost and more rapidly deployed; these attributes allow for a much wider range of people and organizations to build spacecraft. While small satellite platforms are not nearly as capable as their larger, more "traditional" counterparts, they facilitate large growth and investment. Since just 2015 well over 600 CubeSats have flown (Swartwout 2019a; 2019b) and it is expected that much greater adoption of the small satellite form factors will continue with investments on the order of tens of billions of dollars (Market Reports 2019; Research and Markets 2019). The schedule and cost savings appear, so far, to justify the reduced capability imposed by this smaller form factor.

With the growing interest and investment in these platforms there is a growing level of scrutiny being applied to the small satellite industry. Common space industry practices are being applied to small satellites that have been developed for larger one-of-a-kind space assets ("Design, Construction, and Testing Requirements for One of a Kind Space Equipment" 1986; U.S. Department of Defense 1986; Johnson-Roth 2011). Essentially, many organizations are attempting to develop small satellites to the Class D or (the ambiguous) sub-Class D level of system engineering and mission assurance.

While Class D missions can be applicable to any size of space system, the reality is that small satellites generally do not meet the intent of Class D. The growing prevalence of small satellites is also starting to violate the assumptions that Class D was predicated on: that these are one-of-a-kind. Class D is a higher risk posture but has evolved to (or always did) assume a relatively high probability of mission success. The small satellite community, and the design principles therein, have evolved from the concept of pushing the boundary on faster innovation. The small satellite community's innovation cycle was enabled by the community adoption of the containerized 1U standard. This standard has since been adapted to larger form factors but the fundamental design trades were developed in a form factor that were amenable for widespread adoption. This standard has allowed the community to focus on innovation in processes and platform capabilities atypical of larger scale missions.

Further, these systems are greatly constrained and often are not capable of achieving something like Class D. The form factor imposes many physics-based limitations (volume, mass, power), many technologies are relatively low Technical Readiness Level (TRL), and the greater space industry holds many misperceptions about these vehicles (e.g., 50% of all small satellites are dead on arrival to orbit; the actual number is more like 17% (Swartwout 2019a; 2019b). Because of the perception that these spacecrafts are cheaper and faster, their schedules and budgets are often more static than the traditional "big space" paradigm. This drives capability, system engineering processes, and mission assurance.

It is recognized within the small satellite community that high levels of system engineering and mission assurance processes can reduce the innovative intention of small satellites. Where possible, the idea that a small satellite mission will "fit the box" instead of "building the box" has been utilized to help scope missions implemented in a small satellite form factor, as shown in Figure 1. While these ideas have been in the small satellite community for years, they have only recently been more directly discussed (Jasper et al. 2018; Johnson et al. 2018; Tolmasoff, Delos, & Venturini 2017).

In *constraint-driven design*, schedule, cost, and existing limitations (both technical and policy) drive the mission scope and execution plan. This is, so far, how most small satellite platforms have been designed, in contrast to the "big space" requirements-driven paradigm. *Requirements-driven design* prioritizes mission scope over schedule, cost, or other limitations that may drive larger development efforts.

In order to be constraint-driven and reap the benefits of faster and cheaper, a mission's scope must be well defined and limited (Tolmasoff, Delos, & Venturini 2017) or the scope must be flexible to reductions as constraints are realized. This idea can be challenging and even abrasive to much of "big space," but it is familiar to many small satellite crafters (Swartwout 2019b). Assuming this step can be taken with mission stakeholders, the next most important attribute to a constraint-driven mission is scaling the systems engineering practices: the focus of this paper.

It is a common refrain for those working on small satellites that certain practices are not conducted "because it's a SmallSat." This is, in of itself, not sufficient or technically correct. Small satellites go through all of the same phases and steps as any space vehicle however there are many practices and processes that are either done on a very small scale or are not applicable. The processes also tend to be iterative versus serial, with smaller scale processes happening throughout the mission lifecycle. Tailoring of these practices and processes to be constraint-driven is discussed and recommendations are made based upon experience from various AFRL programs and the University NanoSatellite Program. Further, good practices for improving the resilience/robustness of space vehicles, without necessarily increasing system engineering or mission assurance burdens, are discussed.

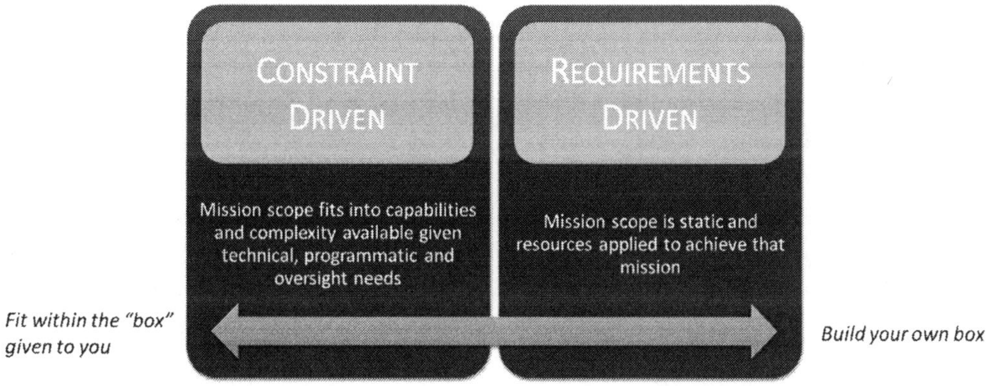

Figure 1: Constraint vs. Requirements driven missions (Jasper et al. 2018).

Vitality of Mission Scope

Unlike the wide and deep requirements of the traditional spacecraft development approach, the requirements in a constraint-driven model are kept at a high level and focused on the specific capability that is required to be demonstrated on orbit. The scope should cover the overall definition of what the mission is supposed to accomplish and a specific description of what the end result should be. Detailed (or deep) requirements are still necessary, but they are only created when they are needed. It is important through this process to not over-define the solution space but rather the problem that needs to be solved.

The key piece of information here that drives scope is the minimum viable product, which is tied directly to the capability that should be validated on-orbit. Each capability has at least one on orbit demonstration associated with it for on-orbit validation purposes. Note that if there is not an on-obit test associated with the capability then the associated development is de-scoped from the mission. This scoping effort drives many of the systems engineering design trades that are determined through a mission lifecycle and fundamentally bound the programmatic constraints of the mission.

In constraint-driven models, the scope of a mission is controlled, not fixed. It is expected that the scope will change throughout the mission lifecycle; this change is documented throughout mission lifecycle at programmatic reviews. It is critical that programmatic discipline is maintained to only add capabilities when they have made space by removing other capabilities first. Scope creep, where mission stakeholders add desired capabilities outside the necessity of the minimum viable product, is a real danger to the success of a mission.

It is critical to document exactly how and when a mission's objectives are to be achieved by showing the major products, milestones, activities, and resources required for the mission. In traditional management, the scope, cost, and schedule imply high quality attributes that are locked down at the start of the project; conversely, in a constraint-driven model, the mission should deliver the desired scope, in the time allowed, within the budget allocated, and to the quality aspired to. The systems engineering processes tailoring therefore is a conversation between all stakeholders, which is clearly defined at the beginning of a mission, so that mission expectations and programmatic constraints can be realized as early as possible in the mission lifecycle.

Systems Engineering Processes

Although there are several definitions of the various systems engineering processes in use today, this paper references the IEEE 15288 definitions and process breakdown. Table 1 presents the processes that we discuss in this paper, broken down into technical management processes and technical processes, following the breakdown given in the DOD's "Best Practices for Using Systems Engineering Standards" (Office of the Deputy Assistant Secretary of Defense 2018). Note that several of the processes called out in 15288 are considered out of scope for this paper, consisting of acquisition, supply, life cycle model management, infrastructure management, portfolio management, human resources management, quality management, and knowledge management. While critical to the success of an organization, this paper neglects a discussion of the larger processes and focuses on the processes that are within the scope of a single project.

Technical Management Processes

Accurate project planning is generally considered the most difficult task that systems engineers are assigned. One often quoted rule of thumb is to multiply your most accurate cost and schedule estimate by pi (3.14) to get a realistic estimate, or the constant e (2.72), if you are feeling optimistic. While there are always unknowns that will trip up any program plan, there needs to be recognition that there are significant outside factors that drive this perception. One significant one is the inherent optimism that is required when making a program plan under competitive circumstances. A green-light schedule that assumes zero problems will always be unrealistic, especially under cost-plus contracting; firm-fixed price contracting has a strong tendency to bring clear-eyed realism to cost and schedule discussions, with those most familiar with the challenges of the project being able to inject their concerns into the planning process. This, in turn, forces difficult discussions significantly earlier in

Table 1. Systems engineering processes

Technical Management Processes	Technical Processes
• Project Planning • Project Assessment and Control • Decision Management • Risk Management • Information Management • Configuration Management • Quality Assurance • Measurement	• Mission Analysis • User Requirements Definition • System Requirements Definition • Architecture Definition • Design Definition • System Analysis • Implementation • Integration • Verification • Transition • Validation • Operation • Maintenance • Disposal

the program, requiring more realistic cost-benefit trades to be made at the user level, and helps temper unrealistic expectations from mission sponsors. Cost overruns are still a significant fact of life, but when constraints imposed on missions are rooted in reality and cancellation is more than a threat, and a valid option for a program, cost and schedule realism can become part of the organizational culture.

Generally, even the cheapest missions will still undergo the full review process that is inherent in the project assessment and control process. Tailoring is applied to the individual review, with a certain level of informality and relaxation of rigor to the requirements that are levied at each review. One critical piece that is shared between this and the decision management process is to push the decision-making power as far down the organization as possible (Jasper et al. 2018). This has the effect of minimizing the need to bring the reviewers up to speed on the current state of the mission and allows the review to focus on the current issues that need addressing before moving forward. Continuity of management (driven by short schedules) also helps this process drastically, maintaining familiarity with the mission and knowledge of the previous decisions.

Risk management is generally one area where process is tailored and that generally falls to an identification of the primary risks at every review, with appropriate mitigation as it relates to the mission success. For many missions, large risk items that would be unacceptable for higher-class missions are routinely accepted, such as the use of industrial quality electronics and unknown radiation susceptibility (generally a community practice). Mitigating the lack of more structured risk management is the smaller teams that are enforced by the low budgets of these missions. The improved communication within small teams allows the systems engineers to discover the risks inherent in specific courses of action. Also, key is having the available expertise that is necessary to understand newfound risks and mitigations quickly.

The adoption of new toolsets, such as Confluence or other wiki-based systems, has enabled significantly lower friction information management processes than predecessor file-based toolsets. Accompanying delegation down the organization structure of approval and review authority and relaxation of some of the related formalisms also simplify and speed information transfer through the wiki-based toolsets.

Configuration management and quality assurance are often lumped together because of the overlap in objectives and processes. A significant relaxation that is applied is the ability to work both tests and assembly procedures without detailed procedures. When the test requirements and test flow have been discussed with the appropriate approvers, the test can be run and documented live, providing a significant speedup. Integration with the wiki-based information management system has also improved the ability to capture critical information from the procedure. Some relaxation of the standard two-person rule has been tolerated, mostly in relaxing the knowledge requirements of the second person, where a tech or engineer with unrelated expertise can review and sign off on an action with appropriate explanation by the acting person. Flight hardware handling practices include ESD safety, smocks, hairnets, gloves, and a class 10K clean environment.

Measurement processes are generally associated with tool location and calibration tracking. Poor calibration practices can come back to damage a spacecraft in the most inopportune times, giving little options to scale back calibration practices. In general, tools lost inside spacecraft hardware can be mission ending, but with spacecraft as small as these, there are few opportunities to misplace tools.

For many of these systems engineering processes there is the recognition that while process can improve consistency, it can also reduce individual responsibility and ownership. Delegation of authority is critical to improve ownership and responsibility such that relaxation of process can reduce cost and schedule without catastrophic results.

Technical Processes

The technical processes in Table 1 generally follow a mission flow, with the exception of the system analysis process, which crosscuts throughout the mission. Figure 2 shows the connection between the processes and the mission lifecycle for a satellite mission.

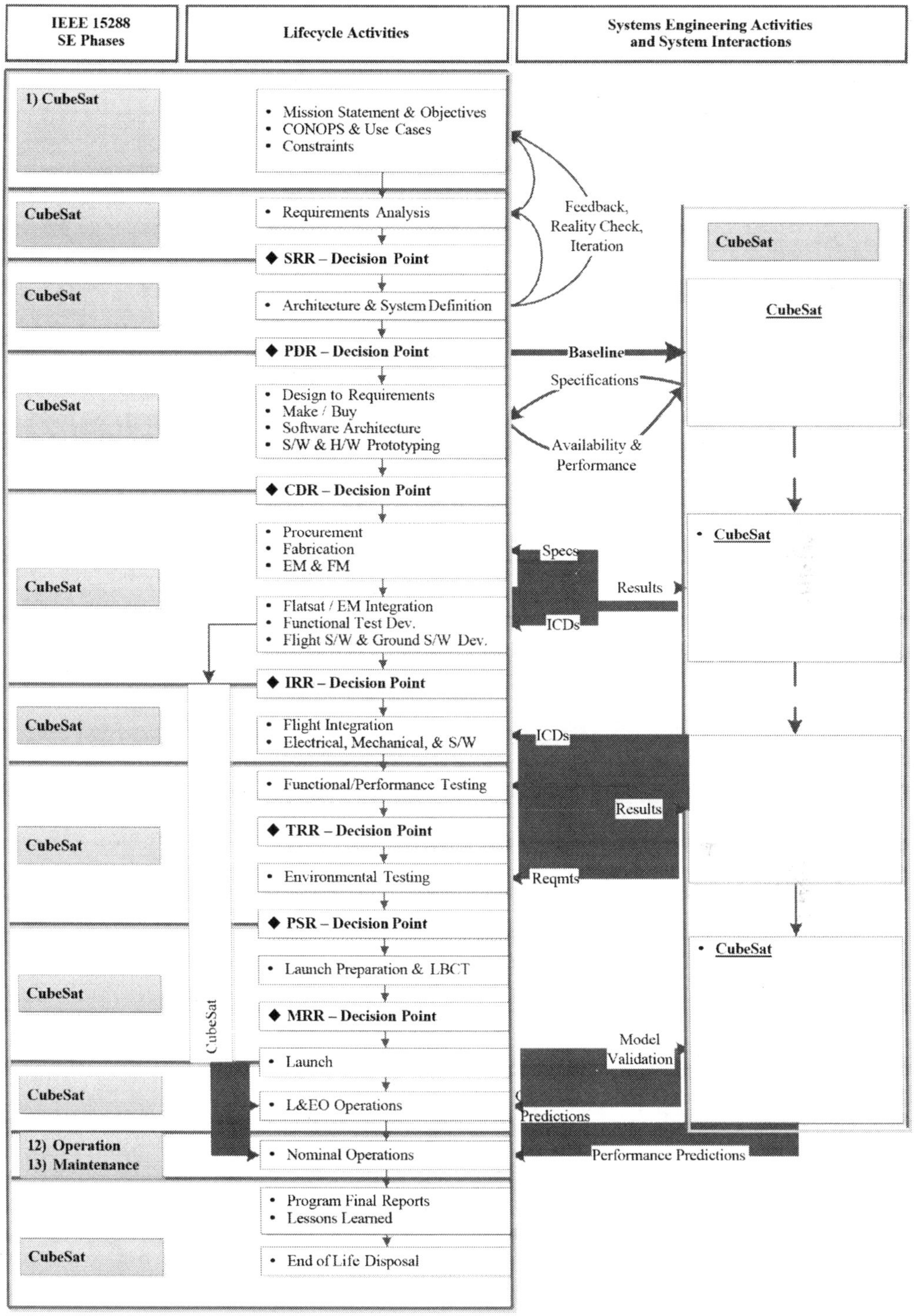

Figure 2. Systems engineering processes and the mission lifecycle.

Concept development

The early stages of mission lifecycle are likely the least well defined. The goal of the early stages of mission development is to identify a self-consistent set of mission objectives, requirements, and architecture that is feasible within cost and schedule constraints. Sometimes this is straightforward, such as when a customer approaches with a well-scoped component test idea. Usually there will be several iterations of concept development, including cost and schedule estimates, returning to the customer to discuss options and possibilities, before a commitment is made.

Concept development generally consists of rapid iterations on the systems budgets, such as communications, power, pointing, navigation, etc., evaluating changes to the mission and experiment CONOPS enabled by various options. Impacts to the requirement set and system architecture guide new decisions. Key performance parameters drive decisions and guide the selection process.

The baseline for the system is implemented and documented in a system design model that consists of the CONOPS; requirements and constraints with functional, performance, and environmental testing defined; architecture; and the performance and cost budgets. Further detail is required in a risk assessment and mitigation plan. The generation of a self-consistent system design model is necessary to progress to PDR.

This top-level description of the concept development process most likely applies across all mission classes. The key idea that changes with mission class is the fact that the process looks for a well-scoped minimum viable product with reasonable agreement that it is worthwhile to embark on for the cost and schedule resources available. This can be low risk, such as a widget testing mission, or high risk, such as attempting to interface with a global satcom constellation.

Design definition

Between PDR and CDR, the design is fleshed out through procuring or developing the subsystems and components that meet the detailed requirements. One simplification applied is a strong preference toward buy-in make/buy decisions. Full design rigor is generally expected when the decision is to make the component in-house.

One simplification to the review and approval process that can be adopted is a peer technical review, which is a detailed, but often informal, assessment of the work conducted on a component, subsystem, system, etc. The intent is to get a second set of eyes to better catch errors or omissions of best practices, to cross-pollinate ideas, and to provide more cross-team communication. This review may come from a subject matter expert or similarly skilled engineer from another project.

In many cases, it is quicker and cheaper to begin prototyping early in the process, allowing engineers to evaluate design and component selection decisions while providing time to correct mistakes. The increasing complexity of the various ICs available today increases the challenge of catching errors at the schematic level; often the only way to determine if a chip can perform the required task is to prototype the circuit and work out the proper settings by hand.

Canonically, software development work prior to CDR should be limited to architecture, prototyping, and planning. However, most missions can attest to the wisdom of an early start to software development. In this case, the use of non-EEE parts can significantly enhance the capability of the processing on the spacecraft and has enabled significant sophistication in the flight software of CubeSat missions. At the same time, if mission scope can be reduced sufficiently, the mission logic may be able to fit in basic microcontrollers, significantly reducing the time and financial investment required for software development.

Certain judicial enhancements to the mission at this design stage can minimize cost and personnel commitments during both testing and operations. One requirement that is generally carried on AFRL/RVEN missions is to be power positive in a tumble. This requirement enables the critical components to be reduced to the power, TT&C, and connecting subsystems (usually command and data handling). The elimination of the attitude determination and control subsystem from the system safe mode allows for both simplified operations (e.g., business hours only, progressing to unattended operations) and a reduction in testing in the ADCS system, due to the knowledge that it is not a critical subsystem.

System analysis

The system design model is the central analysis tool that supports the systems analysis portion of the systems engineering process. The model captures the mission and experiment CONOPS, the requirements flow down, product break down and work break down structure, and the ICDs.

The interaction between the system design model and the system changes throughout the mission lifetime. Most of the design work on the mission occurs in the system design model prior to PDR. Between PDR and CDR, the model is updated to reflect component availability and feasibility, cost benefit analyses, design trades, and evolving schedule and cost constraints. The final CONOPS scenarios, design, and expected performance are captured at CDR.

After CDR, the model serves as the basis for defining test campaigns and incorporating test results into performance predictions. The model informs flatsat, hardware-in-the-loop and software-in-the-loop testing and provides the proper location to incorporate the record as-built performance and calculate system margins and capability. It also provides the ability to analyze the impact of a failed test and informs the decision to modify the design, modify the test, or accept it as-is with a waiver.

In AI&T, the model helps specify the functional and environmental testing to ensure a test-as-you-fly approach. As final testing wraps up the model is used to develop operations plans and a mission planning toolset for use during early, nominal, and contingency operations.

Implementation

The functions of implementation are the procurement and fabrication of the various parts, components, and subsystems. One particularly powerful simplification of this process is the use of a flatsat, where non-flight boards and harnesses are electrically integrated in a tabletop setting. This encourages the rapid identification and correction of flaws in design, ICD mismatches, and most non-mechanical issues. The flatsat allows for breaking connections and breakout box level verification of key measurements that are infeasible after mechanical integration.

The flatsat also allows for early functional test development, which provides time for iteration on the functional test procedures and helps catch design flaws, allowing for later flight hardware functional tests to focus on workmanship flaws. A heavy focus is placed on test scripting.

The flatsat also provides an ideal platform for flight software testing. The acceleration of flight software development on the flatsat is likely sufficient justification for the apparent extra effort even without the other advantages described here.

Assembly, Integration, & Testing (AI&T)

The AI&T phase of any mission can make or breaka mission's schedule and/or budget. During this phase of mission development, many of the investments or shortcomings made in earlier missing phases are realized.

Traditional mission AI&T focuses heavily on carefully developed integration procedures with multiple levels of inspection and may even include the construction of an engineering unit to test these procedures. These practices, while well suited for requirements-driven missions, significantly increase both the cost and schedule for the mission. For constraint-driven missions, similar levels of mission assurance can be achieved through the application of some simple design practices and lean integration processes specifically applied to mission-critical integration activities.

In general, small satellite missions are designed and built by much smaller teams than their traditional counterparts. This allows the design team to also act as the AI&T team. Having these functions so closely coupled allows the AI&T team to become experts with their system during the design and since they do not hand off AI&T to a separate team, there is less need to meticulously design integration procedures. With this level of understanding of the design intent, procedures can focus on critical integration activities, such as optical alignments, and less on the integration of more robust systems.

The testing and verification of constraint-driven missions also vary significantly from the traditional paradigm. While the same objectives of verifying that the system will survive launch and perform the mission objective still apply, the level to which this verification is performed is where constrain-driven missions vary the most. For these missions, it has been found the that the greatest return on investment comes from the following basic tests:

Functional Day-in-the-Life (DITL)

DITL testing, when properly designed, should accurately demonstrate the critical functionality of the spacecraft. This usually focuses first on initial startup and system checkout and then exercises operational modes. Some simple error detection and recovery testing may be performed; however, it is not the intent of constraint-driven DITL to exercise all edge cases, but instead to simply verify that the system performs as intended. This test specifically includes the launch and early operations sequence.

Power Characterization

As the power subsystem represents the lifeblood of the spacecraft, significant efforts are expended to verify the full functionality of the subsystem. This includes verification of the depth of discharge, recharge through solar panels, autonomous recognition of safety limits on the battery, proper inhibit functionality, load testing and switching, and proper telemetry production.

Long-Range Communications Verification

Small satellite systems present a unique opportunity to test a full end-to-end communications path of the satellite that simply could not be performed with larger systems. Due to their size, satellites can either be tested by free air radiating with a significant distance between the test antenna and the satellite or with an actual ground station asset. Many issues can be discovered by performing a long-range test that would otherwise be missed when using an antenna hat or when performing attenuated hardline tests.

Command and Execution Test

Full verification of the software functionality is required, although there is some flexibility on whether that is performed on the flatsat, flight vehicle, or simulator. This is an execution of each command in the Command and Telemetry List (CTL). The depth to which all the various permutations of arguments for each command are verified is allowed to fluctuate depending on the mission.

Full Functional Test

Functional testing on the balance of the subsystems is allowed to stay at a high level, emulating the expected use cases that each component may see in operations. If failures are encountered, further investigation is required. Often there are edge and corner cases that are not well explored or tested, and these can be discovered in orbit. The expectation is that as long as the critical subsystems are well characterized, these faults are recoverable and can be dealt with during operations.

CG/MOI Testing and Polarity Checks

These tests gather the required information to ensure that the ADCS system and algorithms are provided with the most accurate information. The polarity checks also ensure that the sensors and actuators were installed correctly.

Other tests that may be performed, given the specific risk tolerance posture of the mission, these include EMI/EMC testing, detailed ADCS testing, and payload performance testing.

Vibration Testing

More traditional systems may test all components independently prior to integration and modeling the integrated system prior to full vehicle vibration testing. Constraint-driven missions can realize significant cost and schedule savings by only vibration testing the fully integrated system and limiting modal modeling to only extremely sensitive components.

Thermal Vacuum Testing

Testing the system under both hot and cold vacuum ensures that the system will perform as designed in orbit. While the duration and number of cycles can vary from mission to mission, limiting the number of cycles can significantly reduce the cost and schedule.

Mission Operations

Traditional mission operations consist of several operators sending command sets up to the spacecraft in a serial process. This method of controlling is well suited for the requirements-driven mission, as it provides a human in the loop to ensure that the spacecraft remains operational as much as possible.

For a constraint-driven mission, this operations paradigm must also be adjusted. Many of these missions have much more constrained operations budgets that drive a push to operate as "lights out" as possible. "Lights out" operation is a method of operating a spacecraft with either very limited or zero controllers sitting in the mission control center.

This operations method is achieved through the careful design of the constraint-driven system to include two design principles. The first of these is a tumble proof COM link. By providing a communications link that can still close the link with the ground even in a tumble, operators can recover the vehicle from anomalies much more quickly and monitor the system state of health even if it currently unable to recover from a current power condition. Once the power system recovers, operators can then proceed with bringing the system back online.

The second operations enabling principle is to utilize the DITL testing to develop mission operations scripting. By developing and utilizing this scripting during the testing phase, these command sets can be "canned" and used for future operations. By designing in this way, operations planning can then be accomplished during a weekly planning meeting rather than the more traditional daily planning.

Conclusions

The majority of space organizations have evolved to be requirements-driven, such that meeting mission goals and scope takes precedence over cost and schedule, due to the limited access to space. However, as access to space continues to expand for small satellites, and the need for rapid capability development increases, schedule and cost are driving mission lifecycles. These constraint-based missions require tailored systems engineering practices that prioritize demonstrated capability with a lower performance over undemonstrated capability with higher performance. The small satellite community should adopt a process that verifies mission success, allowing mission validation to occur on orbit and allowing rapid demonstration of capability.

References

Jasper, L. E. Z., L. Hunt, D. Voss et al. 2018. "Defining a New Mission Assurance Philosophy for Small Satellites." *SmallSat Conference,* Logan, UT, August 4-9. Paper No. SSC18-WKII-05.

Johnson, M. A., P. Beauchamp, H. Schone et al. 2018. "The Small Satellite Reliability Initiative: A Public-Private Effort Addressing SmallSat Mission Confidence." Paper presented at the *SmallSat Conference,* Logan, UT, August 4–9. Paper No. SSC18-IV- 01

Johnson-Roth, G. 2011. "Mission Assurance Guidelines for A-D Mission Risk Classes." 2011. *Aerospace Corporation,* TOR-2011(8591)-21.

The Market Reports. 2019. "Global Small Satellite Market Insights, Forecast to 2025." January. Report Code: 1237587.

Office of the Deputy Assistant Secretary of Defense. 2017. "Best Practices for using Systems Engineering Standards (ISO/IEC/IEEE 15288, IEEE 15288.1 and IEEE 15288.2) on Contracts for Department of Defense Acquisition Programs." http://www.acq.osd.mil/se/.

Research and Markets. 2019. "Global Prospects for the Small Satellite Market, 2018-2022." March 27.

Swartwout M. 2019a. "CubeSat Database." Saint Louis University. https://sites.google.com/a/slu.edu/swartwout/hom e/cubesat-database

Swartwout, M. 2019b. "CubeSat Mission Success: Are We Getting Better?" In *Proceedings of the CubeSat Developers' Workshop,* CalPoly, April 23.

Tolmasoff, M., R.S. Delos, & C. Venturini. 2017. "Improving Mission Success of CubeSats." In *Proceedings of the U.S. Space Program Mission Assurance Improvement Workshop,* The Boeing Company, El Segundo, CA, June.

U.S. Department of Defense 1989. "Design, Construction, and Testing Requirements for One of a Kind Space Equipment." 1986. SPVT-2016-005, ORIGINAL ED., DOD-HDBK-343.

Will a Global Reliance on Space Technology Inevitably Lead the United States to Conflict?

Iván Gulmesoff
American Public University System

ABSTRACT

This analysis aims to provide an assessment of the emerging global threat to the United States satellite infrastructure. Additionally, the analysis provides an understanding of how the greater reliance on satellite infrastructures around the world will increase the threat. While a satellite can be defined as any object orbiting a planet or a star, for this analysis, a satellite will be defined as a man-made machine sent into orbit for a specific purpose. This analysis will begin with a brief comparative analysis between the current dispute over the Spartly Islands in the South China Sea through the lens of the social dominance theory. The following sections will introduce the current threat to the US satellite infrastructure followed by the policy recommendations.

Keywords: space technology, U.S. conflict, global reliance, global threat, satellite infrastructure

¿Una dependencia global de la tecnología espacial conducirá inevitablemente a los Estados Unidos al conflicto?

RESUMEN

Este análisis tiene como objetivo proporcionar una evaluación de la amenaza global emergente para la infraestructura satelital de los Estados Unidos. Además, el análisis proporciona una comprensión de cómo la mayor dependencia de las infraestructuras satelitales en todo el mundo aumentará la amenaza. Si bien un satélite se puede definir como cualquier objeto que orbita un planeta o una estrella, para este análisis, un satélite se definirá como una máquina artificial enviada a la órbita para un propósito específico. Este análisis comenzará con un breve análisis comparativo entre la disputa actual sobre las Islas Spratly en el Mar del Sur de China a través de la lente de la teoría del dominio social. Las siguientes secciones presentarán la amenaza actual para la infraestructura satelital de EE. UU. Seguida de las recomendaciones de política.

Palabras clave: tecnología espacial, conflicto de EE. UU., dependencia global, amenaza global, infraestructura satelital

全球对空间技术的依赖会必然导致美国发生冲突吗？

摘要

本研究旨在评估美国卫星基础设施面临的新兴全球威胁。此外，本研究就全球加强对卫星基础设施的依赖如何会导致威胁增加进行了解读。尽管卫星能被定义为任何一个围绕行星或恒星转动的物体，但本文出于特定目的将卫星定义为送入轨道的人造机器。本研究一开始透过社会支配理论视角，对当前南海地区的南沙群岛争端进行一项简短的比较分析。随后的部分将介绍美国卫星基础设施当前面临的威胁，然后是政策建议。

关键词：空间技术，美国冲突，全球依赖，全球威胁，卫星基础设施

Introduction

The United States satellite infrastructure continues expanding as the demand for services increases and space technology improves. As Representative Jim Cooper (D-TN) suggests (Harrison, Johnson, and Roberts 2019) "We (U.S.) are almost as dependent on satellites as we are on the sun itself." Regardless of the function/operation, all U.S. space activities remain critical to meet national security objectives as well as projecting U.S. space power (Stockdale, Aughenbaugh, and Boensch 2018). While the expansion and reliance have enhanced our technological capabilities with communications, remote sensing, global positioning/navigation, broadband, and entertainment, it has also exposed vulnerabilities. In 2016 the U.S. had a total of 576 satellites in orbit while China had 181, and Russia had 140 (Johnson-Freese 2016). By the U.S. having the most satellites in orbit in comparison to other states at any given time increases the threat probability by mere exposure. The exposure of more satellites in orbit increases the threat.

Social Dominance Theory

Social dominance theory suggests societies are grouped through various factors while introducing group-based oppression (Pratto, Sidanius, and Levin

2006). The grouping among societies is evident not only by the separation of states worldwide but within each state by political parties. Group-based oppression within states may not only lead to social inequalities but conflict as well. For members of the out-group, the perception of further exclusion may bring them closer together. As Alexander et al. explain, individuals, create a sense of belonging to a social category or group (Alexander, Levin, and Henry 2005). Because they view themselves as members of a specific group their behaviors follow the image of the same social category.

As Pratto et al explain (295, 2006) "According to the social dominance theory, group discrimination tends to be systematic because social ideologies help to coordinate the actions of institutions and individuals." *However, could this ideology of social exclusion apply to different states as well? What contributing factors would further drive social exclusion?*

Using the Spratly Islands to Understand Potential Social Exclusions and Conflict

A brief overview of the disputed Spratly Islands might contribute to an understanding of how social exclusions and possible conflict might be driven by external factors. The Spratly Islands have been critical to the import and export of oil throughout the world (Orban 1995). The reason for this is because of the location of the Spratly Islands. These islands are located in the South China Sea where territorially claims could lead to control of the sea lanes as well. What could be implied by any territorial claim would be a control of the oil imports and exports in this region of the world. Oil being a critical resource because of its worldwide dependence could lead to conflict if disputes surrounding the Spratly Islands remain unresolved.

A dispute could be a result of failed internal and external economic balancing strategies. Internal balancing can be defined by improving economic systems and military efforts within a state while external balancing includes forming alliances (Advincula 2015). Forming external alliances could remain critical in many ways. Forming alliances with states with greater internal and external economic balancing could not only benefit members of the alliances but also achieve greater objectives. One of these objectives could be economic growth or military capabilities.

Lack of International Objectives Among Space-Faring States Resulting in Power Projection

While global state satellite infrastructures do not provide a critical resource such as oil through trade routes, however, they do provide critical functions throughout the world that many have become dependent on. The increasing space activities performed by the U.S. can be perceived differently depending on state governance and alliances. Space activities could be considered either a threat or beneficial to multiple states throughout the world. These views, as well as their implications, will be further discussed. As Aganaba-Jeanty explains (Aganaba-Jeanty 2016) "space activities are judged as either threats to or consistent with space sustainability, rather than as part of articulating the content of space sustainability." In other words, there is no consistency in the space domain in terms of international objectives among the spacefaring states. The space domain previously considered a sanctuary is now at the forefront of global threats and policy implications. By each state aiming to achieve its objectives in space, they are also presenting a threat of power projection. As states increase their space presence sooner or later there will be an inevitable conflict over the benefits of supremacy (Handberg 2017). As Orban suggests in regards to the Spratly Islands (74, 1995) "these anxieties are well-grounded in reality when one considers the fact that China is building its forces at a time when there is no obvious major threat to its militarily or territorially." Taking this example into consideration, one might wonder what the influence and control of one country over the majority of satellite infrastructures might lead to—possible conflicts and disputes?

The first threat of power projection in space occurred in 1957 with the orbital flight of Sputnik 1 (Handberg 2017). At the time, the act of orbital flight alone could have easily been considered a breach of national airspace because it crossed several states in transit. The United States and the U.S.S.R. had both begun projecting their technological prowess of space. In 1962 the United States detonated the 1.4 megaton Starfish Prime hydrogen bomb at an altitude of 248 miles that disabled six critical satellites from the United Kingdom, the U.S. and the U.S.S.R. (Steer 2017). However, what could have been considered a threat of military coercion, quickly abated with the ramification of the U.S. space policy that later developed into the Outer Space Treaty. The Outer Space Treaty defined space as a sanctuary. Today, the threats to our satellite infrastructure may come in the form of intentional damage or destruction of a U.S. commercial or government satellite in orbit. The threat to the U.S. satellite infrastructure can be divided into two most likely variables: nefarious acts of state or non-state actors. The spacefaring states with the most advanced space technology capabilities include Russia with the implementation of the Aerospace Force and China with its Rocket Force (Johnson-Freese 2016). On the other hand, non-state actors pose a completely different threat and challenge to the U.S. satellite infrastructure.

The challenge and threat of non-state actors on the satellite infrastructure are significantly different because their goals may be inspired by political or economic beliefs (Johnson-Freese 2016). While the first threat of power projection with the Sputnik 1 was mitigated through diplomacy, the method a state government will negotiate with a non-state actor may be completely different. Johnson-Freese provides two examples of protentional NGO threats in the future; (8, 2016) "an activist billionaire wanting to promote transparency could deploy a constellation of satellites to monitor and then tweet the movements of troops worldwide" or "criminal syndicates could use satellites to monitor patterns of law enforcement to elude capture, or a junta could use them to track rivals after a coup."

ASAT Systems

The levels of vulnerability can be determined by the ASAT system capability successfully achieved by a state or non-state actor. However, a significant factor to the vulnerability of the U.S. satellite infrastructure is a state or non-state actor's ability to access the low earth orbit (LEO) environment. The vulnerability of U.S. satellite infrastructure is limited only to states and non-state actors with the technology to access LEO. The cost of space access is defined by all expenditures required for the launch infrastructure, launch operations, and the payload size necessary to operate in space (Stockdale, Aughenbaugh, and Boensch 2018). ASAT systems can operate from either land-based or space-based environments. ASAT weapons are either kinetic (physical striking) or non-kinetic (such as lasers, high-powered microwaves or electromagnetic pulses) (Harrison, Johnson, and Roberts 2019) However, both environments have their challenges to effectively operate ASAT systems.

In the space environment, the primary concern with any ASAT system will be space debris. As Chow suggests, within the next decade there will be an estimated 16,000 additional satellites introduced into the LEO (Chow 2018). These additional satellites will not only create a congested environment but contribute to the growing space debris in orbit. NASA scientist J.C Liou suggests, for any sustained space operations, the removal of space debris must begin by 2020 with an average removal rate of five large objects per year (such as rocket bodies or unusable satellite parts) (Chow 2018). Not only could the U.S. satellite infrastructure become vulnerable to space debris but also accept the risk of collision, damage, or a misunderstanding considered a hostile act from the introduction of new orbiting satellites. This method of negotiation may reflect different variables not considered with standard negotiations with state actors. Stockdale et al. explain how introducing large constellations of small satellites would require monitoring, tracking, operation, and management to prevent confusion (Stockdale, Aughenbaugh, and Boensch 2018). With the U.S. already having the most satellites in LEO, space debris and the growing congestion will compound the threat of vulnerability.

Congestion creates another challenge and threat of vulnerability. Using modern technology an ASAT system can remain undetected for extended periods in orbit. As Chow explains, ASAT weapons could be developed as stalkers to follow in close proximity to US satellites (Chow 2018). By having the ability to sustain orbit and maneuverability, a US satellite could be severely damaged or destroyed by the ASAT. This threat of vulnerability from a stalker weapon could be compounded by a lack of situational awareness of a stalker while remaining defenseless. Similar to a stalker weapon used in space that can remain undetected until used, two comparable ground-based threats to vulnerability exist as well.

Ground-launched ASATs and directed energy technology remain a significant threat to the U.S. satellite infrastructure. As Chow suggests, ground-launched intercontinental ballistic missiles (ICBM) with a range above 2,000 km can be considered a threat to any satellite (Chow 2018). As previously mentioned, most satellites are placed in LEO, by having ICBM systems with this capability they would not only pose a threat to satellites in LEO but also in MEO. While ICBM systems remain a critical threat, directed-energy weapons such as lasers have also been developed to either disable or destroy satellites (Set 2015). While directed energy weapons have not been used to destroy a satellite to date, they remain a significant threat as the technology continues to improve.

Like most technologically advanced systems, the U.S. satellite infrastructure is also vulnerable to cyber-attacks. One example of a significant cyber-attack explained by Anderson and Sadjadpour when (Anderson and Sadjadpour 2018) "an Indian threat actor throughout 2015 to 2016 repeatedly created phony corporate websites for Oshkosh Corporation, an American defense company, to capture credentials from its private internal business network, and continued to target aviation companies, including jet engine manufacturers and satellite companies." Not only was this an example of a significant threat exposing U.S. vulnerabilities, but it also proved cyber-attacks can spread throughout multiple industries regardless of the state boundaries. Any cyber-attack on the U.S. satellite infrastructure would potentially be directed towards the transmission of radio frequency signals that affects launch systems, maneuverability, tracking or recovery systems. Each cyber-attack would not only affect operations but introduce significant costs in repairs as well as the implementation of new cybersecurity measures to prevent reoccurrence.

The first ASAT technology was introduced by the U.S. two years after the successful deployment of the Sputnick and since then has expanded to state and non-state actors (Harrison, Johnson, and Roberts 2019). As no non-state actors with capable technology have emerged to this day, this identification of operators will be narrowed down to the most likely state ASAT capable adversaries of the U.S with the most reliance on their satellite infrastructure. Every state has its own reliance and limitations on different aspects of its satellite infrastructure which has turned the space environment into an offense-dominant domain (Kopeć 2019). This dependence on satellite capabilities has also increased the risk of new ASAT systems to protect those assets. This section identifying operators and capabilities will begin with the most likely and technologically advanced adversary and end with the least likely.

China and New Space Technologies

China began developing new space capabilities to include hit-to-kill ASAT technology in the 1980s following the introduction of the Reagan administration's Strategic Defense Initiative (SDI)

(Johnson-Freese 2016). While the SDI, nicknamed Star Wars, was intended to establish an anti-ballistic missile defense system, it prompted China to develop its space technology to keep up with the space technology of the United States. Contrary to the United States striving to achieve defensive space capabilities, China focused on offensive capabilities. In 2006 China successfully tested disruptive ASAT technology by firing lasers at U.S. satellites and in 2007, hit-to-kill ASAT technology was achieved destroying a defective weather satellite (Johnson-Freese 2016; Chandrashekar 2016). In addition to these kinetic ASAT capabilities, China also improved its anti-ballistic missile capability. These improved capabilities have been proven to target satellites beyond LEO in the geosynchronous earth orbit (GEO) (Chandrashekar 2016). China has not only developed and successfully tested kinetic ASAT technology but non-kinetic technology as well.

In addition to the direct-ascent kinetic-kill vehicles, co-orbital satellites, China has focused its technology on developing directed energy weapons and jammers (Pollpeter 2016). While these newly developed weapons could be used strictly for space ASAT they have a much greater purpose with a national security strategy in mind. China seeks to develop and improve its technology with the Yaogan constellation of satellites to include electronic intelligence (ELINT), high-resolution optical sensors, and synthetic aperture radars (SAR) (Chandrashekar 2016). As a larger strategic goal, China aims to use this advanced technology to deter U.S. seventh fleet operations in the Indo-Pacific region.

Russia and New Space Technologies

Russia became a larger threat to the U.S. satellite infrastructure with the introduction of the Aerospace Force. As Vice President Mike Pence announced while presenting the U.S. space reorganization initiative (Harrison, Johnson, and Roberts 2019) "Russia has been designing an airborne laser to disrupt our space-based systems. And it claims to be developing missiles that can be launched from an aircraft mid-flight to destroy American satellites." While the United States remains a major space power in regards to satellite technology, Russia's continued advancements in space technology continue as a threat to the U.S. satellite infrastructure. The most recent Russian Federation military doctrine includes new technological advancements prioritized to monitor orbiting objects as a means of averting military conflicts (Deák 2015).

For the United States to maintain its space dominance it will have to respond to actions that threaten safe satellite operations.

Iran & North Korea New Space Technologies

In regards to capability, Iran and North Korea are a lesser threat than China and Russia. Iran and North Korea continue struggling with the capabilities necessary for the successful deployment of a satellite in orbit. Although Iran formed an Iranian Space Agency in 2003, it has only been able to successfully launch a few satellites into LEO (Harrison, Johnson, and Roberts 2019). However, Iran has successfully developed a non-kinetic ASAT capability. In 2003, Iran successfully jammed a commercial satellite from a location in Cuba (Cesul 2014). Like Iran, North Korea also has a high failure rate of successful orbital satellite launches. While Iran and North Korea may not maintain a successful capability to launch a satellite into orbit, they both continue developing ASAT capabilities that can pose a threat to the U.S. satellite infrastructure. North Korea has proven and continues developing ICBMs with increased range while Iran could develop a crude direct-ascent ASAT capability with its current ICBM systems on an unguided warhead (Harrison, Johnson, and Roberts 2019). While both Iran and North Korea do not pose the same threat in regards to capability as China and Russia, they both continue developing ground-launched ASATs that could be a threat in the future.

China, Russia, Iran, and North Korea: Their Respective Approaches to Space Objectives

Russia's space doctrine is part of its military doctrine with the intent of providing continued orbital surveillance while establishing space-based weapons (Deák 2015; Harrison, Johnson, and Roberts 2019). Unlike other spacefaring states, until recently, Russia maintained the record of the most successful orbital launches. This consistent record alone increases the measured intent. Furthermore, Russia was the first state to form a Space Force to reinforce its military doctrine. The Russia Space Force has developed several ASAT capabilities since its establishment. One ASAT capability is a modification of a MIG-41BM aircraft to deliver and launch a missile that could reach a satellite in LEO (Harrison, Johnson, and Roberts 2019). While this ASAT capability on a MIG-41BM aircraft has not been successfully proven it does raise the concern of its potential. Russia has several ASAT systems that have been proven successful. The period between 1963 and 1982 twenty satellites were destroyed using Russia's ASAT technology and in 2018 Russia completed its seventh test of a new PL-19/Nudol direct ascent ASAT system (Harrison, Johnson, and Roberts 2019). By assessing Russia's space doctrine to its current and past achievements towards this goal Russia is assessed as being the number 1 threat to the U.S. satellite infrastructure.

China's approach to space objectives is part of a larger, more strategic goal. By having the Yaogan constellation of satellites with specific systems to monitor operations in the U.S. Seventh Fleet area of operations the data provided is intended to enhance China's anti-ship ballistic missile system (Chandrashekar 2016). China's motives for space operations are tied to a larger goal—to dominate the Asian-Pacific region. The military buildup of the Spratly Islands in the South China Sea is one example of China's long-term goal. As Stashwick suggests (Stashwick 2018) "China's extensive island-building projects in the Spratly Islands, the aggressive harassment tactics of its maritime law enforcement and paramilitary fleets, and its rejection of binding arbitration rulings on both those activities threaten the rules-based international order and pose political, economic, and potentially military threats to U.S. interest in the region." By comparing China's current and past operations in the South China Sea to meet their objectives to their current space operations the measured threat intent to the U.S. satellite infrastructure should remain high/number 2.

Iran's objective in space is (26, 2019) "to deny the United States the ability to use space in a regional conflict is critical to its security" while North Korea's space doctrine remains unknown. While both states aim to achieve space objectives their failed attempts of orbital flight and lack of proven ASAT technology keep them from being as much of a threat to the U.S. satellite infrastructure as other states. However, both states have proven successful cyber-attacks that could affect a satellite in orbit by non-kinetic means. As Crosston suggests, frequent and successful cyber-attacks have come from North Korea and Iran (Crosston 2017). While some may see both Iran and North Korea less threatening in terms of capability, their intent may lead them to improve their existing cyber-attack technology as a future non-kinetic ASAT system.

Policy Appraisal Suggested Pathway Build

The benefits of satellite technology have proliferated among spacefaring states creating congestion in LEO requiring an avenue of long-term peaceful stability of space to prevent conflict. As Secretary of State Hillary Clinton explained in 2012 "a code of conduct will help maintain the long-term sustainability, safety, stability, and security of space by establishing guidelines for the responsible use of space." While the threat was as apparent then as it is now in 2019, no code of conduct was ever established. Secretary of State Hillary Clinton was not only trying to establish "red lines" but also implies there is a need to change existing policies with the introduction of space weaponry and the growing use by spacefaring states. However, one might wonder, where do we draw the red lines in space?

As previously mentioned with the comparison of the disputed Spratly Islands and the measured intent of China, the space domain remains questionable as a global commons and jurisdiction. The Outer Space Treaty defines space as a global commons while international law reinforces the same argument. International Law states the high seas, the atmosphere, Antarctica, and outer space shall (6, 2016) "lie outside the political reach of any one nation-state." But what happens when many more states want to reap the space benefits only a few spacefaring states have enjoyed since post World War II? As Johnson-Freese suggests "more actors wanting and expecting to exercise that right inherently creates more opportunities for clashes among space actors than in the past when space was the exclusive purview of only a small number of states." With modern-day technology, many states throughout the world rely on space technology for military and civilian uses.

The oldest space treaty to date is the Outer Space Treaty of 1967 that was intended as an agreement for the safe exploration of outer space to include the moon and other celestial bodies.

Cesul explains how one of the primary purposes of this treaty was for the space environment to be used for the good of mankind (Cesul 2014). However, the treaty has fallen short of preventing the militarization of space and the introduction of space weaponry. Article 3 of the Outer Space Treaty states that the (6, 2016) "establishment of military bases, installations and fortifications, the testing of any type of weapons and the conduct of military maneuvers on celestial bodies shall be forbidden." As previously mentioned above, with the threats and capabilities of each state, there has been a blatant disregard for this treaty. To date, the treaty consists of 105 states that have completed ratification (Balleste 2017). Space weaponry such as ASAT continues being used today regardless of the Outer Space Treaty of 1967. A significant issue today are the inconspicuous aspects of space technology were satellite systems can be developed with added ASAT systems (Johnson-Freese 2016). Modern-day satellites require multiple systems to operate. A system designed for ASAT could easily be introduced in addition to a satellite intended function. While the original intent of the treaty served its initial purpose, it slowly diminished with the continued introduction and dependence of more satellites and ASAT technology.

Although the Outer Space Treaty intended to encompass all activities with the peaceful use of space to include the moon and celestial bodies, it was later combined with four additional treaties. The General Assembly combined the 1967 Outer Space Treaty with the 1968 Rescue Agreement, the 1972 Liability Convention, and the 1975 Registration Convention (Leib 2015). Each of these four treaties added what was regarded as essential requirements to the Outer Space Treaty. The rescue agreement was for the safe return of astronauts or space objects, the liability convention established guidelines for damage caused by objects in space, and the registration convention recorded objects launched into orbit.

A major concern with this treaty has been in regards to jurisdiction. Article 1 of the Outer Space Treaty explains how (1042, 2017) "outer space, including the moon and other celestial bodies, shall be free for exploration and use by all states without discrimination of any kind, on a basis of equality and in accordance with international law." This article implies the jurisdiction is specific to areas of space and not suborbital vehicles (Balleste 2017). With the introduction and continued development of commercial space operations, this jurisdiction becomes questionable. As Steer explains, commercial actors in space are vital to providing services to both civilian and military technology demands (Steer 2017). As previously mentioned, a non-state actor could potentially establish a constellation of satellites. If these were used for malicious intent damaging a nearby satellite, where would the jurisdiction fall? On the closet celestial body?

The Outer Space Treaty achieved its goals in some ways but not in others. As Johnson Freese suggests space was militarized during World War II even though article 3 of the Outer Space Treaty states space should only be used for "peaceful purposes" (Johnson-Freese 2016). During World War II, space was used to allow missiles to travel longer distances with less effort than it would require in lower altitudes. Following World War II space technology was greatly enhanced for the U.S. with the help of German scientists who came to the United States and later designed the V-2 rocket for human space flight and the first ICBMs (Sariak 2017). Even though several states signed and ratified the Outer Space Treaty it appears some only abide when it is convenient and aligned with national security objectives.

The Outer Space Treaty also fell short of achieving its goals because of the blatant disregard of space weaponry and ASAT technology. However, this blatant disregard may be part of a larger national security strategy by U.S adversaries. Set explains how (189, 2015) "outer space essentially is an effective force multiplier and combat on Earth shall become increasingly dependent on the developments on the final frontier in varying degrees as testified by successive military operations and doctrines since Gulf War I." While no weaponized platforms have been placed in space, the intent by states who signed and ratified to the Outer Space Treaty exists. The Space Threat Assessment of 2019 indicates Iran intends to weaponize space and Russia and North Korea continue using non-kinetic ASAT technology against the U.S. and allied forces (Harrison, Johnson, and Roberts 2019).

In several ways, the policy addressed the initial concern and threat following World War II. As Handburg suggests, the Outer Space Treaty provided a multilateral agreement among spacefaring sates to continue space exploration without unnecessary confrontations using and accessing LEO (Handberg 2017). This understanding not only advanced the exploration of space but created dialogue for spacefaring states to work together for a common purpose. The stability created an environment where multiple states could work together for the same benefits (Johnson-Freese 2016). However, while the original problem may have been addressed to an extent, the problem is slowly evolving with the introduction of modern technology and more spacefaring states.

As of 2016 sixty states have a space program or are actively conducting space operations (da Costa 2016). This is a significant increase from the elite few spacefaring states that established the Outer Space Treaty of 1967. One might wonder, with the continued benefits and reliance on space technology, have all sixty states ratified and abided by the Outer Space Treaty of 1967 and the 1968 Rescue Agreement, the 1972 Liability Convention, and the 1975 Registration Convention? As previously mentioned, there have been many questionable instances of weaponization and arguments of jurisdiction and sovereignty. However, a significant attempt to challenge the Outer Space Treaty came from a group of unlikely states. Brazil, Colombia, Congo, Ecuador, Indonesia, Kenya, Uganda, and Zaire proposed the Bogotá Declaration claiming sovereignty over geosynchronous orbit (Leib 2015). Even though the United Nations are any spacefaring state that took this attempt seriously, it was still regarded as a significant attempt to change and disregard the Outer Space Treaty of 1967.

The costs and consequences of the Outer Space Treaty of 1967 go hand in hand. The consequences of not abiding in the Outer Space Treaty will most likely result in a cost of some kind. For example, the use of an in-orbit ASAT to destroy a critical satellite operated by another state may result in some form of retaliation or mediation. Kopeć explains (121, 2019) "deterrence is to convince the adversary that the costs and risks associated with a given activity will outweigh the benefits, and thus, these activities will seem unprofitable to the opponent." Although ASAT systems have been used in space by several states causing enormous amounts of debris, no conflict in space has occurred. While spacefaring states may have ASAT technology, the risks seem to outweigh the costs. Not only would an attack have significant costs, but also question the integrity and intent of the attacking spacefaring state. The result of an attack in space may lead to unexpected consequences by other multilateral agreements (such as the rescue agreement).

Conclusion

In conclusion, the existing threat to the U.S. satellite infrastructure must be carefully evaluated considering all spacefaring state and non-state actors regardless of being an ally or not.

The U.S. dependence on the satellite infrastructure places the risk too high to afford any form of system degradation. China is evaluated as being the #1 threat in terms of capability because of its continued spacefaring achievements. As Ferrera-Snyman stated when discussing capability, (490, 2015) "China in January 2007 shocked the international community by performing an Ant-Satellite (ASAT) test which generated a vast amount of space debris in low earth orbit." On the other hand, Russia is evaluated as the #1 in terms of intent not only because of their spacefaring achievements over a longer period than most states but their goal of establishing space-based weapons that are clearly against the Outer Space Treaty.

Considering the continued introduction of new satellites every year by space-faring states, the accumulation of space debris as well as modern ASAT technology the Outer Space Policy should be revisited and amended. Throughout this process, however, the United States government must consider how their space program is perceived by other states. As Johnson explains, the United States should portray a positive image, and always be considered as upholding the right and responsible use of space for all (Johnson-Freese 2016). This perception would not only assist with cooperating and negotiating with other spacefaring states but also uphold the initial intent of the Outer Space Treaty—maintaining space as a global commons. While one may not be able to accurately predict a conflict in space, a global dependence may lead to similar disputes over sovereignty and international laws such as the freedom of navigation around the Spratly Islands in the South China Sea. As Harrison et al. suggest if we (U.S.) were engaged in a conflict in space (4, 2019) "without our satellites, we would have a hard time regrouping and fighting back. We may not even know who had attacked us, only that we were deaf, dumb, blind and impotent."

References

Advincula, Julian V. 2015. "China's Leadership Transition and the Future of US-China Relations: Insights from the Spratly Islands Case." Journal of Chinese Political Science 20 (1): 51–65. https://doi.org/10.1007/s11366-014-9327-x.

Aganaba-Jeanty, Timiebi. 2016. "Space Sustainability and the Freedom of Outer Space." Astropolitics 14 (1): 1–19. https://doi.org/10.1080/14777622.2016.1148463.

Alexander, Michele G., Shana Levin, and P. J. Henry. 2005. "Image Theory, Social Identity, and Social Dominance: Structural Characteristics and Individual Motives Underlying International Images." Political Psychology 26 (1): 27–45. https://doi.org/10.1111/j.1467-9221.2005.00408.x.

Anderson, Collin, and Karim Sadjadpour. 2018. "Known Unknowns: State Cyber Operations, Cyber Warfare, and the Jus Ad Bellum." Carnegie Endowment for International Peace, 48.

Balleste, Roy. 2017. "World Apart: The Legal Challenges of Suborbital Flights in Outer Space." International Law and Politics, October.

Cesul, Dr B T. 2014. "A Global Space Control Strategy," 17.

Chandrashekar, S. 2016. "Space, War, and Deterrence: A Strategy for India." Astropolitics 14 (2–3): 135–57. https://doi.org/10.1080/14777622.2016.1244747.

Chow, Brian G. 2018. "Space Arms Control: A Hybrid Approach," 27.

Costa, De Leon Petta Gomes da. 2016. "Chinese Geopolitics: Space Program Cooperation among China, Brazil, and Russia." Astropolitics 14 (1): 90–98. https://doi.org/10.1080/14777622.2016.1148465.

Crosston, Matthew. 2017. "The Millenials' War: Dilemmas of Network Dependency in Today's Military." Defense & Security Analysis 33 (2): 94–105. https://doi.org/10.1080/14751798.2017.1310699.

Deák, József. 2015. "Russia's Space Defence from Its Beginning to the Present Time." National University of Public Service, 15.

Ferreira-Snyman, A. 2015. "Selected Legal Challenges Relating to the Military Use of Outer Space, with Specific Reference to Article IV of the Outer Space Treaty." Potchefstroom Electronic Law Journal/Potchefstroomse Elektroniese Regsblad 18 (3): 487. https://doi.org/10.4314/pelj.v18i3.02.

Handberg, Roger. 2017. "Is Space War Imminent? Exploring the Possibility." Comparative Strategy 36 (5): 413–25. https://doi.org/10.1080/01495933.2017.1379832.

Harrison, Todd, Kaitlyn Johnson, and Thomas Roberts. 2019. "Space Threat Assessment 2019." A Report of the CSIS Aerospace Security Project.

Johnson-Freese, Joan. 2016. "A Space Mission Force for the Global Commons of Space." SAIS Review of International Affairs 36 (2): 5–13. https://doi.org/10.1353/sais.2016.0016.

Kopeć, Rafał. 2019. "Space Deterrence: In Search of a 'Magical Formula.'" Space Policy 47 (February): 121–29. https://doi.org/10.1016/j.spacepol.2018.10.003.

Leib, Karl. 2015. "State Sovereignty in Space: Current Models and Possible Futures." Astropolitics 13 (1): 1–24. https://doi.org/10.1080/14777622.2015.1015112.

Orban, J R. 1995. "THE SOVEREIGNTY DISPUTE OVER THE SPRATLY ISLANDS: THE LIMITATIONS OF INTERNATIONAL LAW IN RESOLVING THE DISPUTE," 122.

Pollpeter, Kevin. 2016. "Space, the New Domain: Space Operations and Chinese Military Reforms." Journal of Strategic Studies 39 (5–6): 709–27. https://doi.org/10.1080/01402390.2016.1219946.

Pratto, Felicia, Jim Sidanius, and Shana Levin. 2006. "Social Dominance Theory and the Dynamics of Intergroup Relations: Taking Stock and Looking Forward." European Review of Social Psychology 17 (1): 271–320. https://doi.org/10.1080/10463280601055772.

Sariak, George. 2017. "Between a Rocket and a Hard Place: Military Space Technology and Stability in International Relations." Astropolitics 15 (1): 51–64. https://doi.org/10.1080/14777622.2017.1288509.

Set, Shounak. 2015. "The International Relations of Outer Space: Changes, Continuities, and Contextualities." Jadavpur Journal of International Relations 19 (2): 184–92. https://doi.org/10.1177/0973598415627903.

Stashwick, Steven. 2018. "Getting Serious about Strategy in the South China Sea." Naval War College: Autumn 2018 Full Issue, 7.

Steer, Cassandra. 2017. "Global Commons, Cosmic Commons: Implications of Military and Security Uses of Outer

Space." Georgetown Journal of International Affairs 18 (1): 9–16. https://doi.org/10.1353/gia.2017.0003.

Stockdale, Philip, Scott Aughenbaugh, and J Boensch. 2018. "Low-Cost Access to Space: Military Opportunities and Challenges," 17.

Disaggregating the United States Military: An Analysis of the Current Organizational and Management Structure of US National Security Policy as it Relates to Military Operations in Space

Joseph Myles Zeman
Missouri State University

Abstract

This article was written to provide the reader with a comprehensive assessment of the realities of the current organizational and management structure of US national security policy as it relates to the conduct of military operations in space. To create an encompassing argument, this article considers the current organizational structure of US space policy while acknowledging that space has, in fact, become a warfighting domain. A reorganization of this magnitude has the potential to generate a succinct chain of command for military space operations while condensing the space acquisitions process and ultimately providing military space operations with the attention and resources needed to keep America and its allies safe. However, this article examines whether reconfiguring the current organizational and management structure of US national security space components does, in fact, have the power to accomplish such objectives. This article relies heavily upon the testimonies and documentation derived from both the Department of Defense (DoD) and the US Congress. In addition, it is acknowledged that US policymakers have turned this into a largely bureaucratic and inherently politicized issue. This article ultimately concludes that some degree of reconfiguration to the current organizational and management structure of US policy as it relates to military operations in space has the potential to positively affect the national security space establishment.

Keywords: space policy, organizational and management structure, deterrence, national security, space components, strategy

Desglose del ejército de los Estados Unidos: Un análisis de la estructura organizativa y administrativa actual de la política de seguridad nacional de los EE. UU. En relación con las operaciones militares en el espacio

Resumen

Este artículo fue escrito para proporcionar al lector una evaluación exhaustiva sobre las realidades de la estructura organizativa y administrativa actual de la política de seguridad nacional de los Estados Unidos en lo que se refiere a la conducción de operaciones militares en el espacio. Para crear un argumento abarcador, este artículo considera la estructura organizativa actual de la política espacial de los Estados Unidos mientras reconoce que el espacio, de hecho, se ha convertido en un dominio de guerra. Una reorganización de esta magnitud tiene el potencial de generar una cadena de comando sucinta para las operaciones espaciales militares al tiempo que condensa el proceso de adquisiciones espaciales y, en última instancia, brinda a las operaciones espaciales militares la atención y los recursos necesarios para mantener a Estados Unidos y sus aliados a salvo. Sin embargo, este artículo examina si la reconfiguración de la estructura organizativa y administrativa actual de los componentes del espacio de seguridad nacional de los Estados Unidos tiene, de hecho, el poder para lograr tales objetivos. Este artículo se basa en gran medida en los testimonios y la documentación derivados tanto del Departamento de Defensa como del Congreso de los Estados Unidos. Además, se reconoce que los formuladores de políticas de EE. UU. Han llevado este problema a convertirse en uno que es en gran medida burocrático e inherentemente politizado. Este artículo finalmente concluye que cierto grado de reconfiguración de la estructura organizativa y administrativa actual de la política de los Estados Unidos en lo que se refiere a las operaciones militares en el espacio tiene el potencial de afectar positivamente el establecimiento espacial de la seguridad nacional.

Palabras clave: política espacial, estructura organizativa y de gestión, disuasión, seguridad nacional, componentes espaciales, estrategia

解析美国军事：一项有关美国国防政策（与太空军事行动相关）目前的组织结构和管理结构的分析

摘要

本文目的是就当前美国国防政策的组织结构和管理结构现状为读者提供一个全面的评估，因为国防政策与太空军事行动准则相联系。为提出一个概括性的主张，本文衡量了当前美国太空政策的组织结构，同时承认太空已实际上成为一个战争区域。对该范围进行重新组织，将可能为军事太空行动创造一个简明的指令链，同时压缩太空采购过程，并最终为军事太空行动提供所需的关注和资源，以保护美国及其盟友。然而，本文检验了重新配置当前美国国防空间组成部分的组织结构和管理结构是否真正能完成这类目标。本文（内容）大量依赖由国防部和美国国会提供的证据和文件。此外，公认的是，美国决策者已推动该议题成为一个在很大程度上具有官僚性质和固有政治性的议题。本文结论认为，对当前有关于太空军事行动的美国政策的组织结构和管理结构进行一定程度的重新配置，有可能对国防空间建立产生积极影响。

关键词：空间政策，组织和管理结构，威慑，国家安全，空间组成部分，战略

Introduction

This is one of the most critical times in our national security space history—it will be seen as a strategic inflection point.

—General John J Raymond (April 17, 2018)

At a National Security Council (NSC) meeting on June 18, 2018, the forty-fifth President of the United States of America, Donald J. Trump, publicly directed the Department of Defense (DoD) to begin creating a sixth US military branch, a Space Force. At this same NSC meeting, President Trump leaned on General Joseph Dunford, Chairman of the Joint Chiefs of Staff, asserting, "if you would carry that assignment out, I would be very greatly honored" (Insinna and Mehta 2018). Despite the June announcement being somewhat of a surprise to the majority of the American population, the discussion of a separate "space branch" gained momentum in March 2018, during President Trump's visit to Marine Corps Air Station Miramar in California (Hart 2018). The decision to reconfigure the current organizational and management structure of US national security policy as it relates to military operations in space has faced pushback by high-ranking civilian and military officials. Former astronaut, John Kelly, asserted, "This is a dumb idea. The Air Force does this already. That is their job. What's next, we move submarines to the 7th branch and call it the 'under-the-sea-force?'" (Mosher 2018).

Even though opposition to President Trump's call for a sixth military branch is apparent, it seems that there is no better time to begin assessing the current organizational and management structure of US national security policy as it relates to military space operations. Throughout this article, an analysis is provided to assess whether the current organizational and management construct of US policy related to military operations in space is adequate for addressing the advanced threat from China, Russia, and other adversarial nations. Despite section 1601 of the National Defense Authorization Act (NDAA) for Fiscal Year (FY) 19 demanding that "With the advice and assistance of the Chairman of the Joint Chiefs of Staff, the President, through the Secretary of Defense shall establish under the US Strategic Command (USSTRATCOM) a subordinate unified command to be known as the United States Space Command (in this section referred to as 'space command') for carrying out joint space warfighting operations," the idea has faced opposition. There must be an appropriate degree of analysis for whether this will benefit the current organizational and management structure of US national security space components.

Therefore, this article covers five areas for gauging the level of benefit. These include the maturation of space organizations related to US national security, the case for space security, defining a threat, policy positions, and lastly, policy recommendations.

This article concludes by providing a summary of the recommendations provided within throughout this work. This article notes that the current administration's idea of reorganizing national security space components has come to fruition. In conclusion, this article acknowledges the unfortunate and highly politicized landscape of the national security space enterprise.

The Maturation of Space Organizations Related to US National Security

We are the best in the world at space. Period.

—Lt. Gen. John Thompson

There is no argument that US space operations were founded by extraordinary individuals with an overwhelming fervor to keep US citizens safe from the Soviet threat. The US space story has both civilian and governmental achievements in research and development (R&D), primarily involving missile defense, satellite reconnaissance, and human space exploration. On account of cold war competition and some newly recognized aspirations for space superiority, both the US and the Soviet Union expended time and resources during an era when each nation was recuperating from the terrors of WWII. Although the two countries fought alongside one another to suppress the axis powers of WWII, a shift occurred that put the two nations in opposition. A change in ideology initiated the western hemisphere's fight to contain communism and began the United States' struggle with the Soviet Union. Missile technology developed by German physicists and the subsequent proliferation of their research was also a primary concern for the United States. Not only was missile technology at the forefront of the debate, but also a quickly developing nuclear capability and ever-expanding missile capabilities demanded both militaristic and academically influenced strategic thought. At a time when the destructive possibilities of weapons were seemingly endless and great power competition was the norm, the US began exploring ways to mitigate the Soviet threat.

The United States' ability to alleviate a portion of the Soviet threat came in the form of space-based missile defense, classified satellite functions, and space-based intelligence collection methods. This grand story involves the US Air Force (USAF), Army, Navy, and intelligence community, specifically the National Reconnaissance Office (NRO). Through trial and error, reorganization and disaggregation, adequate funding and lack thereof, the structure of US military space operations began to take shape. Examined below are the organizations that played a role in the execution of this process. Each organization is initially examined from a historical point of view, followed by an analysis of its more recent organization and management constructs. This section is by no means a comprehensive history of the entire US national security space enterprise, but rather introduces the argument from a historical perspective.

The National Security Act of 1947

Signed on July 26, 1947, by President Harry S Truman, the National Security Act of 1947 reorganized and modernized US armed forces, foreign policy, and the intelligence community. Not only did this act cause a certain level of reorganization, but it also formed many institutions that the US government would soon begin to utilize (Central Intelligence Agency [CIA] 2008). The act established the NSC, established the CIA, merged the War and Navy Departments into the DoD, and most importantly, it reorganized the Army Air Corps into an independent Air Force (CIA 2008). The "Declaration of Policy" of the 1947 Act, or Section 2 [50 USC. 401], states that "each military department shall be separately organized under its own secretary and shall function under the direction, authority, and control of the secretary of defense ... to provide for their unified direction under civilian control of the secretary of defense ... and provide for the establishment of unified or specified combatant commands, and a clear and direct line of command to such command" (DNI 1947). Section 2 of the 1947 National

Security Act dictates that the USAF will operate similarly to the other branches, by reporting directly to the newly created Office of the Secretary of Defense (OSD). An aspect of the 1947 National Security Act that is directly relevant to the central argument was the effect this reorganization had on the entire national military establishment, including the various other national security-related departments and agencies. Like the 9/11 Commission Report, a key goal of the 1947 National Security Act was to clarify lines of communication and promote a more transparent culture within the defense department. Nearly five decades before the devastating 9/11 attacks, the US government was reconfiguring its structure with the goal of preventing unclear lines of communication, something that was addressed in both the 9/11 Commission Report and the Intelligence Reform and Terrorism Prevention Act of 2004.

While the contributions of the 1947 National Security Act were monumental for the intelligence community, it was also clear to President Truman that the US must confront a lack of resources for a domain that was becoming a key aspect in the warfighting equation: airpower.

The similarities drawn between the creation of the USAF and the possible creation of another military branch, the Space Force, requires comparison. As noted in detail above, the US typically begins to address its inadequacy after realizing that it has no other choice. The correlation between the construction of the USAF in 1947 and the potential to create a sixth military branch devoted to military space operations in 2019, or soon after, clearly suggests that a symmetrical level of activity could occur. Congruent to how the end of WWII marked the creation of the 1947 National Security Act, the long-lasting conflict in the Middle East, and a re-emergence of great power competition between China and Russia are driving the US to consider how it can adequately address these threats. Even though the intricacies of creating a new military branch or reinstating a singular unified combatant command for space is not laid out in this section, these scenarios are addressed later in this work.

The United States Air Force

Before the USAF was created in 1945, the Air Force Scientific Advisory Group had noted that both long-range rockets and satellites were a "possibility" (Sturm 1967). Fast-forward to the early 1950s and a power competition between two countries, the United States and the Soviet Union, began to drive space policy. Beginning as a study between the RAND Corporation and the Air Research and Development Command, Project 409-40 "Satellite Component Study" soon to be renamed the more infamously recognized Weapons System 117 (WS-117L) program, was created. At a time when intelligence and deterrence were pre-eminent warfighting tools, attaining US governmental cooperation in space remained an uphill battle. In the spring of 1957, Major General David D. Bradburn recounted working on the WS-117L program and brings insight to this argument.

WS-117L was a program that provided the Strategic Air Command (SAC), a tactical air command responsible for addressing the Soviet threat through United States airpower, with reconnaissance satellites. Maj. Gen. Bradburn noted, "the project moved ahead slowly for a lack of money. Then in October, the Soviet Sputnik went into orbit and suddenly there was money all around" (Bradburn n.d., 61).[1] Not coincidentally, the threat drove the Eisenhower Administration into action, akin to the actions taken by presidential administrations of the twenty-first century. This same threat led to the creation of the NRO, an intelligence-based entity that engulfed a large number of USAF satellite reconnaissance programs.

However, between the 1960s and 1970s, the US was involved in what would soon become known as one of the most controversial wars of the twentieth century, the Vietnam War. Justified as a means for preventing the spread of communism, the Vietnam War was essentially a proxy between two great powers, the United States and the Soviet Union. The United States' intervention in Vietnam required an increase in intelligence collection, specifically in the form of geospatial intelligence (GEOINT). GEOINT happened to be a domain in which the U2 spy plane and various USAF military reconnaissance satellites could contribute. General Jerome O'Malley began to recognize the utility of military aerospace-related assets upon arrival to an operations center in Ton Son Nhut Air Base located in South Vietnam. General O'Malley arrived at the operations center and immediately asked "where did you get these? [referring to geospatial images laid out before him] I just returned from a mission up there getting my butt shot off trying to obtain the same pictures" (Van Inwegen n.d.). Brigadier General Earl. S. Van Inwegen recalls a member of the team stating, "an SR-71 flew over and took them. The crew was not in any harm's way." This statement of protecting the warfighter resonated with General O'Malley to such a degree that it had become the catalyst to his support for military space power (Van Inwegen n.d.). By 1961, the USAF was responsible for approximately 90 percent of US military space operations (Defense Department Directive 5030.18 n.d.). In addition to contributing to the military space domain, the USAF soon became the primary agent for the National Aeronautics and Space Administration (NASA) research and support initiatives (Defense Department Directive 5030.18 n.d.).

However, in March 1961, the Air Force Systems Command (AFSC) was created to address the disaggregated research, development, and acquisitions activities of the USAF. Six years later in July 1967, the Space and Missile Systems Organization (SAMSO) was created to consolidate USAF space and missile defense activities into a single organization. The following year in 1968, the Aerospace Defense Com-

1 Maj. Gen. David D. Bradburn was assigned to the first USAF satellite project, WS-117L. He subsequently held positions of increasing responsibility in USAF space programs, including Director of Space Systems in Washington and Director of the Office of Special Projects in Los Angeles.

mand (ADC or ADCOM) was the primary entity responsible for monitoring missile warning operations for the Air Force. However, on October 1, 1979, ADCOM was removed from service. The end of ADCOM was the result of a space policy study named "the Navaho Chart" (Defense Department Directive 5030.18 n.d.). The conclusion of the Navaho Chart led Brigadier General James Creedon to begin exploring the possible elimination of both ADCOM and the North American Aerospace Defense Command (NORAD) (Fogleman n.d.).[2] However, the study's conclusions ultimately led to the decommissioning of ADCOM, leaving NORAD unharmed.

In 1982, upon years of studies lead by both senior USAF leadership and the US Congress, AFSC merged with ADC, later known as the Tactical Air Command, to form the "Space Command." Not only did the US Congress criticize the unorganized structure of USAF space activities, but also a January 1982 Government Accountability Office (GAO) report concluded that the DoD must establish a Consolidated Space Operations Center. The report also demanded that the headquarters of a potential Space Command or Space Force should be considered (Van Inwegen n.d., 141). In 1985, the "Space Command" was renamed and merged into the Air Force Space Command (AFSPC), so that it would not be confused with the newly created unified combatant command, the United States Space Command (USSPACECOM). The consolidation of the ADC, SAC, AFSC, and the Air Force Communications Command, was the first step in solving the decentralized structure of national security space operations within the USAF. The USSPACECOM was a direct product of President Reagan's 1983 Strategic Defense Initiative (SDI), and an attempt to more appropriately distribute military space operations among the various military branches. AFSPC would remain the primary USAF entity serving USSPACECOM until it was decommissioned in 2002, following the creation of US Northern Command (USNORTHCOM).

Like the creation of many other great military organizations, the AFSPC experienced trial and error. The USAF, contrary to popular belief, began its use of space-based systems far before the early 1980s. Space assets found their way into the strategic argument soon after the conclusion of WWII. Not dissimilar to the majority of opinions today, the idea of utilizing space assets during the late 1940s was a foreign concept, even to those within the Air Force Chain of Command (Hall and Neufeld 1998, 140). General Ronald R. Fogleman, Chief of Staff of the United States Air Force, describes his understanding of USAF space assets as "embarrassing." General Fogleman also asserts that a lack of fundamental space knowledge was widespread throughout

2 General Ronald R. Fogleman was chief of staff of the USAF (1994–1997). The general graduated from the USAF Academy in 1963. Prior to becoming chief of staff, he was commander in chief of the United States Transportation Command and commander of the USAF's Air Mobility Command.

the entire USAF, and recounted, "Other servicemembers did not know about or understand them because they failed to recognize or take the time to learn just what they could do for the warfighter" (Fogleman n.d.). Not only was the USAF space mission seen as secondary to that of pilots, ground forces, and missile combat crew members, a universal lack of knowledge on space operations added to this common misperception.

The US Army

On October 3, 1957, the US Army formally took its spot in history with its creation of the Redstone Anti-Missile Missile System Office (RAMMSO) in Alabama (Cutshaw 2017). Stemming from the first successful launch of a Soviet Intercontinental Ballistic Missile (ICBM) in August 1957, the Army Ordinance Corps created RAMMSO, an office that would be responsible for the R&D of US missile defense and space capabilities. On April 11, 1958, a mere seven months after RAMMSO was named an independent organization, it was deactivated. Like many other military space organizations, RAMMSO immediately began to experience growing pains. This maturation eventually led to the organization's consolidation into the Army Rocket and Guided Missile Agency (ARGMA) (Lang 2015). While the primary mission of ARGMA was to field an anti-ICBM missile, known as the "Nike Zeus," ARGMA's capabilities were limited due to a December 1956 presidential directive (Lang 2015). Even though creating an anti-ICBM missile was an immediate necessity, the 1956 presidential directive enacted a constraint on how far the Nike Zeus would be allowed to travel (Wade n.d.). Not surprisingly, following the Soviet Union's successful launch of Sputnik, the Nike Zeus' range constraint of 200 miles was removed (Wade n.d.). As a result of the successful Soviet ICBM launch, the DoD authorized the Army Ballistic Missile Agency (ABMA) to launch a satellite into space, giving the US Army the ability to claim that they were "the first in space" among US government agencies.

On December 11, 1961, in a reorganizing trend that continued to repeat itself, ARGMA and ABMA were no longer considered separate organizations but were directed to merge their personnel and functions into the Army Ordnance Missile Command (AOMC) Headquarters (US Army n.d.a). A short time after the merger, the newly formed AOMC moved to the NASA Marshall Space Flight Center, a space research organization created during the Eisenhower administration (US Army n.d.a). Relocation to the NASA Marshall Space Flight center delayed operations and reprioritized some of the Army's brightest space scientists, and ultimately disaggregated Army space operations. This move to NASA between 1958 and 1961 hindered the US Army's space efforts, just like it did the US Navy's. Despite a consistent waxing and waning scenario of the Army's space and missile defense operations, on July 19, 1962, the AOMC executed a successful intercept of a mock ICBM with a Zeus Missile interceptor (PBS n.d.). Coming just in time for the October 1962 Cuban missile crisis, the Army's space and missile defense operations proved to be a vital aspect of the national security of the United States.

Not to be confused with ABMA, the Army's Advanced Ballistic Missile Defense Agency (ABMDA) was created on March 4, 1968. Initially formed out of a project between the Advanced Research Projects Agency (ARPA), and the Nike –X Project Office, the follow-on to the Nike Zeus Anti-ICBM program, ABMDA was directed to provide technical assistance to the Army advanced ballistic missile defense program (Watkins 2018). Until ABMDA's termination in 1974, many of its functions would be given to the Ballistic Missile Defense Advanced Technology Center or the Ballistic Missile Defense Systems Command, both of which would soon be condensed into the current US Army Space and Missile Defense Command/Army Forces Strategic Command (USASMDC/ARSTRAT), more commonly known as SMDC (PBS n.d.). However, the Vietnam War severely diverted the Army's focus and funding from space and missile defense to small arms and field ammunition developments (Boehm n.d.). Between 1977 and 1992, US Army space efforts began to reemerge as things like the Tactical Exploitation of National Capabilities (TENCAP), the Airland Battle Doctrine, and President Reagan's SDI surfaced (Mitchell 1991, 61–65). TENCAP, a program deriving from a 1977 congressional directive, was designed to utilize, where applicable, pre-existing national strategic satellite systems to support Army corps commanders and Naval commanders during theater operations (Mitchell 1991, 72). In addition, the Airland Battle Doctrine stressed the importance of having Army control over military space operations through real-time sensors for addressing the enemy threat (Mitchell 1991, 74–75). Not dissimilar from today, a 1985 report, the "Army Space Initiative Study," provided policy suggestions that enhanced the US Army's use of space (Mitchell 1991, 74–75). Many of the recommendations laid out within the 1985 report were implemented and made positive contributions to the US Army's overall space efforts (PBS n.d.).

Throughout its lifetime, Army space and missile defense operations were reorganized, renamed, and reconfigured many times over. In 1997, the USASMDC/ARSTRAT became the primary Army component providing space and missile defense capabilities to USSTRATCOM. The SMDC's current mission is to conduct space and missile defense operations and provide planning, integration, control, and coordination of Army forces and capabilities to support USSTRATCOM missions like strategic deterrence, integrated missile defense, and space operations (National Research Council [NRC] 2005; US Army n.d.b).

The US Navy

Following the conclusion of WWII, both the US Army and the USAF took on the challenge of researching and developing missile defense technologies. With pressing threats emanating from the Soviet Union, it was the US Navy's task to understand space's atmospheric intricacies better. More than any other service, the US Navy began academic-like research of the space domain through the Naval Research Laboratory (NRL), the Johns Hopkins Applied Physics Laboratory (APL), and the Applied Research Laboratory at Pennsylvania State University (Laurie 2001).

In addition to the research being conducted by the US Army, the US Navy was heavily dependent on the V-2 rocket. The V-2 was an extremely unreliable rocket system that the US Army had seized from the German military in the late 1940s. Unlike the Army, the Navy's mission was not missile defense, but rather its task was to explore how the United States could place satellites into orbit for intelligence purposes. Further, it was also the Navy's job to assess the potential for space-based communication capabilities. However, the mission of launching satellites for intelligence collection became a point of contention between the Army, Navy, and USAF. On September 9, 1955, despite disputes regarding which branch was a better fit for the job, the task of satellite "launch" was ultimately awarded to the NRL, dubbed "Project Vanguard." It is important to note that the USAF would have been responsible for conducting this mission, but the need to develop the Atlas rocket, the launch vehicle for ICBMs, took precedence. On March 17, 1958, just two and a half years after the program began, the NRL delivered a 3.5-pound satellite into orbit. This satellite launch kicked off the Naval Space Surveillance System, a program that would remain a cornerstone for naval space operations and become the precursor to "The Fence," a program that is still in operation today (NRC 2005; US Army n.d.c). Following the successful launch of the 3.5-pound "Minitrack" satellite, President Eisenhower began the formation of NASA, a move that would disrupt and divert naval personnel, resources, and research. The creation of NASA was a move that affected not only the US Army and USAF but also the US Navy. Although the Navy began to lose resources and personnel to NASA, the DoD realized the need to harness Navy satellite technology. As a result of this apparent need, on April 10, 1962, the Navy Astronautics Group (NAG) was commissioned to operate the Navy Navigation Satellite System (NNSS), also known as "TRANSIT" (US Archives n.d.). Despite losing a substantial amount of resources to NASA, just as the Army had, the Navy was able to continue executing vital research for satellite systems architecture throughout the 1960s (US Archives n.d.).

A key piece that allowed the Navy to continue its substantive research in space was a revision to DoD Directive 5160.32 in 1970. This revision allowed each of the services to continue developing various satellite systems for navigation, communications, mapping, meteorology, and various other mission sets. For the Navy, the principal entity responsible for this research was NAG. NAG was primarily responsible for monitoring ultrahigh-frequency (UHF) and extremely high frequency (EHF) satellite operations, aspects that are still vital to global positioning satellites (GPS) and communications satellites today. On October 1, 1983, the Naval

Space Command (NAVSPACECOM) was created, subsequently placing NAG under its purview. The relocation of NAG under NAVSPACECOM was an attempt to organize naval space operations more appropriately, an action that repeated itself just two years later. Upon the establishment of USSPACECOM in 1985, NAVSPACECOM began to serve as the primary naval space component to this command ("Naval Space Command" n.d.). In 1990, NAG was renamed the Naval Satellite Operations Center (NAVSOC) (Kennedy and Crawford 1998). However, it was the NAVSPACECOM that maintained the majority of USSPACEOM's workforce and served as the Alternative Space Control Center of USSPACECOM's center located at Cheyenne Mountain Air Force Base, Colorado (Boehm n.d., 23). In conclusion, after the decommissioning of USSAPCECOM, NAVSPACECOM also disappeared. The responsibilities of naval space operation currently fall under the purview of Space and Naval Warfare Systems Command.

US Space Command

On September 23, 1985, USSPACECOM was established at Peterson Air Force Base in Colorado Springs, Colorado. USSPACECOM was responsible for overseeing all US military space operations, including those belonging to the Navy, Army, and Air Force (Roeder 2018). Between 1985 and 2002, USSPACECOM grew substantially, acquiring different mission sets along the way. In 1990, USSPACECOM acquired responsibility for space launch, and soon after in 1993, it gained responsibility for ICBMs (Roeder 2018). Throughout this time, the command began to inherit and transform many pre-existing facilities into bases, such as the Space Operations Center at Schriever Air Force Base, Patrick Air Force Base, Cape Canaveral Air Force Station, and Buckley Air Force Base (Roeder 2018). Despite the consistent and exponential growth of US military space operations, USSPACECOM was deactivated on October 1, 2002. The deactivation of USSPACECOM was a consequence of the newly established USNORTHCOM, which shifted space operations under the purview of USSTRATCOM, and created the need to redirect resources and attention to the fight against terrorism, de-emphasizing the role of military operations in space. Due to USSPACECOM's deactivation, all military space operations were reconfigured under the USSTRATCOM, transitioning a large majority of the responsibility back to the Air Force's AFSPC. As noted previously, a culture engulfed by reprioritization and reorganization has plagued the US military space community for decades. As we will see later, this trend continues to repeat itself, even today.

National Reconnaissance Office and the Intelligence Community

Established on September 6, 1961, as a classified agency in the DoD, the NRO was a culmination of various military programs that eventually formed the intelligence community's first space organization. Heavily influenced by the Gary Powers U-2 shoot down, President Eisenhower demanded then Secretary of Defense, Thomas Gates, begin exploring options for space intelligence collection. Gaining its notoriety from the Navy GRAB program and the CIA CORONA program, the NRO was established to pursue the intelligence community's most heavily classified satellite programs (Laurie 2001). Both the GRAB and CORONA program derived from the anticipated vulnerability of the U2 spy plane, which was reinforced by the May 1, 1960 shoot-down of Gary Powers over the Soviet Union. Formed by direction of the CIA's Allen Dulles and the DoD's Robert McNamara, the NRO's roles and responsibilities toed a line that had yet to be drawn (Berkowitz 2011). Air Force Under Secretary Joseph Charyk was named the first Director of the NRO (Boehm n.d.). However, before the NRO's establishment, an immediate issue facing the organization was an undetermined leadership structure. In response, there came a series of four "agreements" that ultimately contributed to the NRO's organizational structure. The first agreement established the National Reconnaissance Program, asserting that the United States Intelligence Board would set requirements for the organization (Boehm n.d.).

The second of the agreements established a few of the management and organizational constructs for the entity. On July 23, 1962, Dr. Charyk created what would become the basic organizational structure for the NRO, essentially splitting the organization into four distinct program areas. The four programs were project A, the Air Force's satellite reconnaissance program, project B, the CIA's satellite reconnaissance program, project C, the Navy's NRL program, and project D, a joint USAF, and CIA aerial and reconnaissance program (Boehm n.d.). The structure set forth by Dr. Charyk only reflects how decentralized the NRO was becoming. At its core, the NRO was unlike the Army's AOMC, the Navy's NAG, or the Air Force's ADC; instead, it was an organization that took the best pieces of each branch and oversaw only what was of interest to its goals; collecting intelligence on the Soviet Union and protecting against a nuclear war.

On March 13, 1963, the NRO was formally established as an agency within the Defense Department, but also managed to maintain a Deputy Director position, which would be filled by a CIA official. This agreement allowed both the Director of Central Intelligence (DCI) and the DoD to keep a close eye on NRO operations. Just two years later, on August 11, 1965, a fourth agreement reinstated influence back to the DoD, a decision that removed the requirement of the DCI to play a role in the Deputy Director position. This decision essentially removed any authority the DCI previously had in the decision-making equation and relinquished it to the Secretary of Defense. During the NRO's first decade of life, it would see many "agreements" or compromises between the Secretary of Defense, Robert McNamara, and the DCI. These agreements eventually lead to the DoD's complete authority over NRO operations. However, the trend reversed once again. Primarily resulting from miscommunication, the Secretary of Defense, acting as the lead administrator of the NRO, began to see his attendance at Executive Committee (EXCOM) meetings as unnecessary. He began to send his assistant to the semi-annual EXCOM meetings, leading to his eventual disengagement with the organization. However, the EXCOM meetings were abolished in 1976. In addition to this, executive order 12036 gave the DCI "full and exclusive" authority over the National Foreign Intelligence Program budget, and as a result, the Director of the NRO began to report directly to the DCI, once again changing who was in charge.

In addition to a consistently changing leadership structure, the lines of communication between Congress and the NRO were virtually nonexistent. Classification levels and technical lexicon deterred many members of Congress from even attempting to understand the NRO's role in combating the Soviet threat (Laurie 2001, 19). General knowledge of NRO activities was widely viewed by members of Congress as unnecessary. The U-2 program director and CIA Deputy Director Richard Bissell recalled that "a few members of the Armed Services and Appropriations Committees in each chamber oversaw the activities on the Intelligence Community and virtually all oversight was conducted behind closed doors" (Laurie 2001, 7). This universal lack of knowledge among congressional members was not an issue until the mid-1970s, when the US Congress decided that there should be an increased amount of oversight on the intelligence community's activities, specifically the NRO. Unfortunately, the initiative for increased oversight was conducted by policymakers who were extremely unfamiliar with the NRO's programs. Although congressional oversight of the NRO had increased, even today, there are many members of Congress who are uncertain of the role the NRO plays within the intelligence community.

In addition, a 1989 study titled "NRO Restructure Study" and a 1992 study titled "DCI Task Force on the National Reconnaissance Office" evaluated the current management and organizational structure of the NRO, ultimately concluding that it was inadequate ("Report to the Director of Central Intelligence" n.d.). Despite the highly classified nature of the NRO, it is obvious that the organization suffers from an inability to be publicly understood. This overarching theme spans not just the intelligence community, but also the various military services, as their relationship to the NRO has become hindered due to classification levels.

While restructuring the current organizational and management structure of US national space components is not a novel idea, it is an initiative that transcends the historical context that has been provided. The organizational and historical challenges that US military services have experienced is not only a "space issue," but also an issue that spans across many of the warfighting domains. While the historical perspective provided above contains an abundance of acronyms and instances of reorganization, this is both symbolic and symmetrical to the current US national security space enterprise. In 1983, President Ronald Reagan called for a similar analysis, through the implementation of the SDI. The 1992 reorganization of NRO space assets argued that the organizational and management structure of the institution was inadequate to address the threat. In 2001, Secretary of Defense Donald Rumsfeld led a commission in which he concluded that the US was not prepared to defend its civil and military satellite infrastructure. In 2011, the Obama administration produced the 2011 National Security Space Strategy (NSSS), concluding that space had become increasingly congested, contested, and competitive. And finally, in 2017, the Congress through the NDAA demanded an analysis of the current organizational and management structure of US policy related to military space operations.

The overarching reason for current initiative to reorganize US policy related to military space operations is influenced by the historical context in which these components were brought up. In conclusion, this reorganizing initiative should not be executed simply for the "reorganizational" purposes that support a specific political agenda, but rather it should have positive implications and definitive goals to address the advanced adversarial threat. The bottom-line effect is that space has, in fact, become a warfighting domain (Yiannopolous 2018). This historical perspective must be applied and recounted throughout the latter sections of this article. To begin outlining the realities of a militarized space domain, the next section assesses various deterrence elements for the peaceful use of space and a number of unclassified space capabilities, acknowledges our reliance on these space-based architectures, and ultimately, addresses why we must protect these assets.

The Case for Space Security

> *Dominating in space has now become kitchen table conversation ... and that will benefit this country.*
>
> —Heather Wilson, Secretary of the Air Force (September 27, 2018)

Four Deterrence Models of Space

Released in 2011, under the Obama administration, the NSSS, signed by both the Director of National Intelligence (DNI) and the Secretary of Defense, acknowledged that space transitioned into a warfighting domain. Upon its release, the document acted as the US space policy for not only the Obama administration, but also the ensuing decade. The NSSS took into consideration and built upon the 2010 National Security Strategy, the 2010 National Space Policy, the 2010 Quadrennial Defense Review, and the intelligence community's National Intelligence Strategy ("Fact Sheet: National Security Space Strategy" n.d.). Similar to other US strategic DoD documents, the 2011 NSSS outlined what the environment, objectives, approaches, and challenges are in the space domain. A primary component to the 2011 NSSS is how the US would begin to address the challenges it faced by analyzing how best to mitigate congestion and competition and ensuring success within the contested space domain. To address the contested space environment, the NSSS promotes a multi-layered deterrence approach with the goal of preventing and deterring aggression. The NSSS's multi-layered concept relies on four primary deterrence methods (Stone 2015): 1) deterrence through norms, 2) deterrence through alliances/coalitions, 3) deterrence by denial/resilience, and 4) deterrence through aggression/response ("Fact Sheet: DoD Strategy for Deterrence in Space" n.d.). Each element of this overarching deterrence strategy is outlined below.

Within the NSSS, the first element of the DoD's space deterrence strategy notes that "A broadly-accepted set of international norms of responsible behavior will have positive effects on the safety, stability, and sustainability of the space domain" ("Fact Sheet: DoD Strategy for Deterrence in Space" n.d.). The NSSS elaborates on this point by asserting that even if the reliable US deterrence posture does not single-handedly prevent a bad actor from conducting malicious activities in space, it will at least produce a normative international structure that can identify what is considered malicious and what is not. It is interesting to note that almost eight years after the NSSS was released, US military leaders and policymakers on Capitol Hill are having a similar argument regarding the establishment of international norms in another domain closely related to space: cyber. The strategy for establishing international norms may be academic and inherently theoretical, but it is an argument that proves to be timeless. While succinct and definitive international norms may never be universally accepted, due to the cultural differences of various nations, the discussion is one that must occur at an international level. The establishment of internationally acceptable military and civil space operations may not directly help in mitigating the adversarial threat but may aid in clarifying what is legal and what is not in this technologically advancing space domain. While space continues to become more contested, congested, and competitive, the actions of US allies, and adversaries alike, in the space domain, each have repercussions. The mining of asteroids, the colonization of space, and satellite repair capabilities that possess inherently militaristic and potentially maleficent traits are all examples of actions occurring in space that demand international regulation. What the legality and the proposed solutions to these actions look like is beyond the scope of this article; however, it does raise the issue that action and international agreement must be reached so that the US may remain superior in the space domain.

The second element of the NSSS's deterrence strategy, "Build coalitions to enhance collective security capabilities," mirrors what occurred on the international stage in 1945 ("Fact Sheet: DoD Strategy for Deterrence in Space" n.d.). On June 26, 1945, in San Francisco, California, the United Nations (UN) Charter was signed, becoming operational on October 24, 1945. Chapter 7 of the UN Charter addresses "Action with respect to threats to the peace, breaches, of the peace, and acts of aggression" (UN 1945). Article 42 of the UN Charter states that "members of the United Nations shall join in affording mutual assistance in carrying out the measures decided upon by the Security Council." This article essentially states that if the UN Security Council deems that military action against an adversary is necessary, all UN participants may assist with that mission (UN 1945). Similar to the UN Charter, the second element of the NSSS also supports an idea similar to that of President Eisenhower's push for "massive retaliation." It asserts that "Instead, the aggressor must attack assets and forces of multiple countries, which expands the scope of a conflict and reduces the odds that a potential aggressor can achieve their desired outcome at an acceptable cost" ("Fact Sheet: DoD Strategy for Deterrence in Space" n.d.). This specific strategy highlights the idea of "entanglement," noting that attacking a satellite architecture, whose capabilities and costs are shared among many allied nations, will further complicate the adversary's decision-making calculus, thus increasing risk while simultaneously decreasing the benefit. This concept rests on the notion that the US would be facing a "rational" actor or one that takes into consideration a cost-benefit analysis throughout their decision-making calculus (Delpech 2012).

As a result of coalition building, the second element of the NSSS relies on the idea that UN nations would, in fact, respond to an attack on behalf of their ally, making it less likely for an attack to occur (Schulte 2012, 5).

The third element of the NSSS's deterrence strategy is "Denying the benefit of aggression by enhancing the resilience of space architectures and ensuring that the Joint Force can operate effectively when space capabilities are degraded" (Schulte 2012, 5). This particular element of the deterrence strategy happens to be a topic with numerous strategic implications. In the world of military space operations, the "resiliency" of a specific satellite architecture may be understood as the complete disaggregation of a satellite constellation, the nuclear hardening of a satellite, or even the implementation of "dummy" or non-vital satellite architectures in cohesion with civil satellites to confuse the adversary. Disaggregating certain satellite architectures is a method of reconfiguring certain satellites from "big juicy targets," into architectures that consist of a number of widely distributed mini-satellites (Erwin 2017). The majority of the dialogue regarding military space operations is occurring within the top levels of DoD leadership, and not surprisingly, the USAF produced a white paper stating, "Disaggregation is an innovative opportunity to stay ahead of our adversaries, to change their targeting calculus, and to mitigate the effects of a widespread attack on our space assets" (AFSPC 2013). This USAF white paper also asserts "resilience serves as a deterrent, which may be the best way to preserve our capability by avoiding an attack." A prominent individual supporting the idea of disaggregation is General John J. Hyten, Commander of USSTRATCOM. Gen. Hyten noted that he would discontinue his support for "big juicy targets," asserting, "we are going to go down a different path. And we have to go down that path quickly," referring to a path of satellite disaggregation (Erwin 2017).

In general, the ability to produce nuclear-hardened satellites is a primary feature of creating resilient satellite architectures. The likelihood of an electromagnetic pulse (EMP), nuclear radiation, or any other man-made or natural phenomenon occurring in space directly affects the United States' ability to function terrestrially. Nuclear hardening is a dire component for ensuring the proper functioning of a satellite and is an important component to resiliency. Lastly, the ability to divert the attention of our adversaries by denying and deceiving them with the use of faux or civil satellites is a means of resiliency. Unfortunately, the use of this deterrence strategy comes with a large financial cost and demands a high level of compliance with civilian companies who may be uninterested in participating in the militarization of space. While the cost of this deterrence strategy is great, it possesses the ability to protect critical satellite infrastructures.

Finally, the fourth component of the 2011 NSSS deterrence strategy is deterrence through response. This element supports the idea that should deterrence fail, and an attack on the US or its allies occur, the US would respond, but not necessarily symmetrically. The NSSS states that a response to an attack "may not be limited to the space domain, but rather will occur at the time and place of our choosing" ("Fact Sheet: DoD Strategy for Deterrence in Space" n.d.). Similarly, a previously mentioned initiative created by the Reagan administration in 1983, the SDI, was a program that kept our adversaries, specifically Russia's Mikhail Gorbachev, questioning the true capabilities of US missile defense systems. It was this level of uncertainty that displayed America's ability to effectively deter the Soviet threat. President Reagan's SDI instilled the idea that America's ability to create a strategic defense was so advanced that it would be worthless for US adversaries to execute a first strike against it. The SDI set a tone that the US would not only deny the first strike but would subsequently execute a retaliatory second strike that would be devastating to its adversary. The 2011 NSSS clearly states that should a first strike occur on a US space asset; a second strike would endanger the overall wellbeing of the attacking country. In addition, the 2011 NSSS asserts that should deterrence fail, the US "will use force in a manner that is consistent with longstanding principles of international law, treaties to which the United States is a party, and the inherent right of self-defense" ("National Security Space Strategy" 2011). The strategy utilizes the phrase "in a manner that is consistent" to relay the idea that the US may not respond to an attack symmetrically, but with any means that are "consistent with longstanding principals of international law" ("National Security Space Strategy" 2011). It is the current superiority, in many of the warfighting domains, that allows the US to be decisive when considering which means to use when executing a second strike.

In general, the NSSS lays out the four elements of deterrence that can keep the US and its allies safe from an adversarial attack. Although there have been no recent kinetic altercations occurring in space other than China's anti-satellite test in 2007, there has been a dialogue regarding this possibility. While the US has sought to reorganize US policy related to military operations in space, China and Russia have also prioritized the reorganization of their national security space agencies. The growth of Chinese and Russian space agencies has provoked concern among US policymakers, expressed in various congressional testimonies. US leadership continues to advocate for the reorganization of US policy related to military operations in space so that it may keep pace with the quickly developing organizational constructs and space-based capabilities of Russia and China. It has been argued that the reorganization of US policy related to military operations in space can strengthen the United States' deterrence posture. The reorganization of US policy related to military space operations may improve US deterrence by offering a more streamlined distribution of critical resources and a more efficient use of funding, and it may send an implicit message to its adversaries that space policy and superiority is a national priority.

What We're Protecting—Space-Based Capabilities

For many individuals, space is a mystery. This domain is inherently complicated to understand and widely misunderstood. While, there are astronomers, astrophysicists, and aeronomists, all of whom devote their entire career to the study of space, it is not necessary to assess the intricacies of space here. However, while the complexities of space operations are not the focal point of this article's main argument, there must be a brief explanation of how space works from a non-technical perspective. The most important aspects of this explanation involve orbits and constellations. A constellation is a system of satellites that cohesively work together to accomplish specific goals. There are missions where a single satellite may be enough to accomplish the objective, but often this is not the case. Once satellites are ready for launch, they are subsequently attached to a rocket or launch vehicle and put into a specific orbit or location in space. Satellites are typically located in four primary orbits: low Earth orbit (LEO), medium Earth orbit (MEO), highly elliptical Earth orbit (HEO), or geostationary orbit (GEO). Each orbit allows a satellite to take a path complementary to its mission. LEO is known as the orbit between 600 km and 1,200 km above Earth. LEO is where the United States has, at an unclassified level, conducted the majority of its space operations. LEO is also where the International Space Station (ISS), communications satellites, and GPS reside (M. Williams 2017). One benefit to LEO is its proximity to earth, making it easily accessible. Proximity to Earth is important when certain components need to be repaired or replaced or in instances where humans need to conduct research and return to Earth in a relatively short amount of time. Unfortunately, satellites in LEO experience a great deal of drag due to the gravitational pull of earth. Not only is the gravitational pull a challenge for satellites in LEO, but also is the minimal time a satellite spends over a specific area of the Earth. A satellite in LEO can be expected to travel approximately 28,968 km per hour or faster, orbiting the earth in less than ninety minutes. This means that the satellite is not able to monitor a specific area of the Earth for long periods. This fact makes data collection, imagery, and weather monitoring unappealing for this orbit.

Unlike LEO, MEO is the orbit between 12,000 km and 37,590 km above the earth. One obvious benefit of MEO is that its increased distance from Earth allows a smaller constellation of satellites to monitor the same surface area of LEO. MEO also affords satellites the ability to monitor a specific area of the Earth's surface for a longer period.

HEO is the orbit that is 37,590 km above the Earth. Because the distance of these satellites from earth is so great, they experience less gravitational pull and atmospheric drag and take longer to orbit than satellites in LEO or MEO. The downside of placing a satellite in HEO is the cost associated with getting it to its destination. Not only is the cost challenging, but the logistics of maintaining enough onboard fuel for the satellite to utilize while in orbit is also difficult.

Unlike LEO or MEO, GEO is also designated as a "high Earth orbit"; the difference with GEO is that satellites mirror the orbital time of the Earth, roughly twenty-four hours. While the satellite primarily stays over the same longitude, it may tilt and move either north or south throughout its orbital path. Although the tilting causes some variance in its location, a satellite in GEO allows various organizations in both the DoD and civil entities, like NASA and the National Oceanic Atmospheric Administration (NOAA), to monitor specific areas for longer periods. Listed below, Table 1 addresses the previously examined orbits and their accompanying characteristics. While this article is not designed to address the scientific aspects of space, the primary questions one may ask are what is the satellite's distance from Earth? and how long does it take to orbit? These two components are heavily associated with a satellite's capability and will help guide the reader to more fully understand why specific satellites conduct certain missions. These two questions are by no means a way to fully comprehend the overall capability of a satellite, but they do give the reader a foundational construct to return to throughout this section.

The following section addresses, from a capability standpoint, what the US is protecting in space and why these capabilities need to remain secure. While certain systems and programs will surface throughout this section, it is necessary to lead each section from a capability and mission standpoint rather than from the program-specific name; otherwise, this section becomes inherently polarized and biased towards specific aerospace defense contractors. Within each section below is an analysis of when, how, and why certain capabilities came to be, and what the threats and challenges facing these capabilities are. Throughout each section, it is important to remember that not all of the capabilities mentioned have an inherent militaristic function to them, but they all do, in some aspect, contribute to the US infrastructure.

Table 1. Orbital Specifications

Orbit Name	Orbital Initials	Altitude of Orbit (km)	Details of Orbit
Low Earth Orbit	LEO	200-1200	May rotate around earth in approx. 90 minutes
Medium Earth Orbit	MEO	1200-37590	Experiences less gravitational pull than LEO
High Earth Orbit	HEO	35790	Less gravitational pull, longer orbit times
Geostationary Earth Orbit	GEO	35790 +	Orbits once a day and rotates in the same direction as the earth

Position, Navigation, and Timing (PNT)

Inception of PNT. In 1960, the US Navy initiated a program named "Transit," responsible for all-weather navigation for both military and civilian vessels, but most importantly, for naval submarine navigation (Howell 2018). Transit was a concept that began at the Johns Hopkins APL, and soon became the basis for all future satellite navigation systems. The first launch of a Transit satellite occurred in 1960, and only eight years later by 1968, there was a fully operational constellation of transit satellites (Defense Advance Research Projects Agency n.d.). Upon realizing that position, navigation, and timing capabilities could become a significant contribution to our society; efforts arose to establish the NAVSTAR-GPS, managed by then-Colonel Dr. Brad Parkinson of the USAF Missile Systems Organization (American Institute of Aeronautics and Astronautics [AIAA] n.d.). This program was utilized up until 1996, when the DoD replaced it with the current GPS architecture (AIAA n.d.).

Not Just a Location Service. While many of us today take GPS for granted, it was originally a program created for the DoD; it is now a service that is free to the civilian population. After recognizing GPS's potential benefits for both military and civilian populations, physicists and researchers alike retrofitted the previously created DoD satellites with two primary frequencies, an L1 and an L2 line with a signal on the L1 frequency for civilian use. Even though I cannot provide an exact number of individuals who utilize GPS today, a safe estimation regarding its number of users is in the billions. A common misconception of GPS is that it is primarily used for navigational operations. However, its applications are much more diverse. Other uses of GPS include agriculture, aviation, marine navigation, railroad operations, surveying and mapping, timing, meteorology, public safety, and disaster relief. When disaster strikes, GPS, in correlation with geographic information systems, uses remote sensing technology to formulate maps of disaster areas for subsequent rescue and aid operations. Even though disasters may occur sparingly, the timing function of GPS ensures that communication systems, power grids, financial networks, and nuclear facilities are all precisely synchronized for operational efficiency (AIAA n.d.).

Internal and External Challenges. At Schriever Air Force Base, just ten miles outside of Colorado Springs CO, seven USAF airmen are responsible for maintaining US GPS operations. These highly trained junior officers working for the AFSPC are responsible for the protection of the thirty-one on-orbit GPS satellites that provide GPS functions to billions of users. Colonel John Dorrian, USAF, discussed a classified report called "A Day without Space" during an interview. While omitting any classified information, he conveyed the reports main point by saying, "…the gist of it was that there is no such thing." Referring to a day without space, he stated, "Space capabilities, including GPS, are integrated into everything we do. People count on that capability being there" (Uchill 2016). Not only does the USAF ensure the operational integrity of GPS functionality, but the US Department of Homeland Security has also called GPS "a single point of failure for critical infrastructure" (Resilient Navigation and Timing Foundation 2016). Even though scenarios of complete GPS inoperability are extremely unlikely, instances of GPS intervention are not. It is not uncommon to hear the phrases GPS "jamming" or "spoofing," both of which are words that describe the action of tampering with a GPS receiver.

The consequences of GPS jamming, or spoofing has been recorded by the NASA Aviation Safety Reporting System, reporting ninety or more incidents of GPS jamming around various airfields in less than a year (Buesnel and Holbrow 2017). While these events are typically incidental and are not intended to disrupt the overall functionality of vital GPS systems, they do have negative repercussions. These "jamming" incidents are typically caused by seemingly innocent commercial drivers trying to avoid managerial oversight regarding speed limits and unplanned "fuel brakes." Martin Faga, former Deputy Secretary of the Air Force, said that "People who sell these devices," referring to GPS jammers, "say they only work for a few yards, which presumably is just enough for what the person buying the device is trying to hide" (Uchill 2016). However, as Faga elaborated, "the reality is that most of them jam GPS' for a couple of miles, which creates problems" (Uchill 2016). While jamming may not be the crux of GPS issues, physical threats and potential intervention from an international adversary is a concern. During the same interview, Col. Dorrian spoke about his concerns with space becoming a highly contested domain and that the USAF is considering a change in the AFSPC staffing policy. Likewise, in that same interview, Col. Dorrian also mentioned that the GPS team at Schriever Air Force Base (AFB) constantly rotates its staff of inexperienced officers, who are around the average age of twenty-three (Buesnel and Holbrow 2017). While the ages of these highly capable junior officers is not the central argument, Col. Dorrian felt that it was important enough to mention. Col. Dorrian also noted that human error is, in fact, an inherent aspect to any military endeavor, and that the USN's October 2015 initiative that began teaching its sailors celestial navigation if the GPS infrastructure were to become inoperable, solidified the DoD's concern with GPS reliability. So, while the DoD focuses on mitigating threats to the GPS system from both international adversaries and US citizens, it is simultaneously looking for ways to mitigate internal challenges.

Satellite Communications (SATCOM)

The Birth of SATCOM. By the early 1960s, the DoD began development of a communications satellite program named ADVENT. The program was created by ARPA in conjunction with the US Army and the USAF. Two years into ADVENT, Secretary of Defense Robert McNamara canceled the program on account of high costs, inadequate payload capacity, and an unrealistic launch capability. All of these faults and inconsistencies lead to the creation of the Defense Communications Agency, now the Defense Information Systems Agency (Spires and Sturdevant n.d.). Before ADVENT's discontinuation on August 31, 1962, US Congress signed into law the Communications Satellite Act of 1962. The act was intended to "provide the establishment, ownership, operations, and regulation of a commercial communications satellite system, and for other purposes." The DoD soon realized that SATCOM capabilities would become inherently commercialized. By 1961, NASA awarded AT&T, RCA, and the Hughes Aircraft Company contracts to begin production of space telecommunications satellites. By 1964, the three companies each had two operational communications satellites in orbit, thus placing a level of reliability on commercial capabilities. While the DoD has since created its own set of military communication satellites, the bureaucratic process and fragmented organizational structure of military satellite communications (MILSATCOM) has plagued the process. Nonetheless, programs like MILSTAR, the Advanced Extremely High Frequency (AEHF), UHF, and Wideband Global Satellites (WGS), have taken shape despite the challenging MILSATCOM landscape. Although the DoD satellite architecture has continued to advance US capabilities, specifically US Army capabilities, it still relies upon insecure commercial satellites for a number of its critical space operations.

Dependent on Communication. Put simply, the use of telecommunications satellites that can provide beyond the line-of-sight communications is one of the two most essential satellite applications used in military operations today, the other being GPS (Kusiolek 2010, 6). The speed and mobility of systems like the Navy's Multiple User Objective System, and the capabilities of the WGS, speak for themselves. In addition, capabilities like Nuclear Command and Control and Command and Control

Battle Management Communications are a central tenant of our offensive and defensive nuclear capabilities, each relying on communications. However, if these two systems were to be compromised, it may delay, disrupt, and potentially deny the United States' ability to relay critical information about adversarial activities to the warfighter and those in the nuclear command centers. Since the need for SATCOM is apparent, the need to protect these systems should be equally as obvious. As we will see in the section below, the need to protect MILSATCOM is not the only aspect that must be addressed, but so too is the US dependence on supplemental unprotected commercial SATCOM.

The Challenges of Creating Cohesive Communications. Even though MILSATs have their own systematically derived hardware challenges, this section deviates from these aspects and assesses the issues from a broader perspective. As with most DoD space programs, the ability to phase out legacy satellites while simultaneously deploying new ones is an integral aspect on both the financial and operational fronts. To fully integrate a new constellation of MILSATs, the three previously mentioned segments - terminal, control, and space - must all be reconfigured to adapt to new technologies. Not only is backward compatibility a challenge, but so is the convergence of commercial and military systems. While one organization may surpass the other, typically commercial industry over government, each must create technologies that work in a cohesive manner. It sounds simplistic, but all three segments of SATCOM must be interconnected for them to work. A prime example of this challenge is the new "M-Code" capability. The USG fielded the new WGS program, a currently orbited satellite, but the national security space industry struggled to produce the necessary Earth-based terminal components for this system to become usable. So, while the capability was fielded, there was still the challenge of connecting each component with one another.

Another argument lies within the US Army's overdependence on commercial SATCOM. Major Andrew H. Boyd of the US Army reiterated that a critical threat regarding MILSAT is that "The US Army's most critical vulnerabilities is its overreliance on SATCOM, one which most of its mission command systems depend" (Boyd 2017). As noted before, a key argument of this article is the advanced adversarial threat that the United States faces in space, primarily from Russia and China. While the thought of a threat emanating from within the United States itself may seem far-fetched, Maj. Boyd states that "The increasing need for SATCOM bandwidth has led the US military to channel its operational communications through the leased networks of commercial satellites; these lack adequate protection against jamming and are susceptible to state-actor influence" (Boyd 2017). Maj. Boyd's use of the words "stateactor influence" can be taken in a myriad of ways, but the language surrounding these words suggests that intervention of SATCOM within the commercial industry can come either internally or externally. Regardless of how much the USG attempts to overcome its dependence on commercial SATCOM, there will always be a need to supplement MILSAT bandwidth with commercial providers (Berlocher 2008, 6). It is said that as much as 80 percent of all USG satellite communications traffic, including that of the military, is carried over commercial SATCOM systems (Berlocher 2008, 6). During an interview, Colonel Earl Madison (Ret.), former Chief of Staff for National Security Space Architect (NSSA), responded to what he believed the largest issue facing MILSATCOM was, stating, "There is no overall adult supervision on military space communications; now, reorganizing the current space management structure may not totally fix that, but hell, it's a damn good step in the right direction."[3] In addition, a report on the major policy issues in evolving space operations produced by the Mitchell Institute notes that "Early, clear, and public legal processes to indemnify all commercial and international space services and systems that support national security is essential to building resilient architectures with robust contribution from these [commercial] sectors" (Vedda and Hays 2018).

Space-Based Infrared Monitoring

The Beginning of Infrared. As early as 1948, scientists from the US government began exploring the possibilities of detecting and tracking missiles by their heat signatures. WS-117L became the primary DoD space-based reconnaissance and surveillance program spearheading this mission. By November 1958, the program had morphed into the Missile Defense Alarm System (MIDAS) (Richelson 2007). During the initial testing phase of the space-based infrared detecting capability, individuals within the MIDAS program were primarily interested in producing standalone satellites, with the possibility of successfully detecting a missile launch and or nuclear detonation. After years of trial and error MIDAS morphed into Program 949, then Program 647, and was finally named the Defense Support Program (DSP). A major difference between MIDAS and DSP was that MIDAS orbited in LEO, while DSP orbited in GEO (Richelson 2007). Not only were the satellites' orbital locations a major difference to the program, but so was the fact that the DSP would be the first interconnected constellation of space-based infrared monitoring satellites. The first of the DSP satellites was launched November 5, 1970, with the fourth and final satellite being launched in June of 1989. Between 1979 and 1995, a host of follow-on DSP programs were created and subsequently dismissed due to the conflict between the executive branch and Congress (Richelson 2007). Programs such as the Advanced Warning System, the Boost Surveillance Tracking System, the Follow-On Early Warning System, and the Alert, Locate, and Report Missiles program, were all contri-

3 Derived from a personal interview with Colonel Earl Madison (Ret.). When asked "What do you perceive as the largest issue with military satellite communications and does a reconfiguration of the current space management and organizational construct address this?" this is how he responded.

butions to the dying breed of potential follow-on DSPs (Richelson 2007). But in 1995, the USAF announced its new follow-on program, the Space-Based Infrared System (SBIRS), pronounced "sibbers," a program that is still in use today. After years of use, SBIRS is being reconfigured into what the 2019 NDAA calls the Next Generation Overhead Persistent Infrared Radar or "Next-gen OPIR."

A Dire Component to Missile Defense. Ever since the first V-2 missile launches were conducted by Germany in the early 1940s, maintaining awareness of the ballistic missile threat has remained at the forefront of the US priority list. Even though ballistic missiles were once a primary concern for the US, and remain as such, its focus has begun to shift with the appearance of new weapons systems. Advanced air-launched cruise missiles, tactical nuclear weapons, highly maneuverable hypersonic glide vehicles, and an array of new technologies have demanded that the United States maintain a superior ability to accomplish birth-to-death tracking of these weapons systems. The lexicon surrounding missile defense, missile intercepts, and deterrence sounds like that of the Cold War. However, the major difference in today's landscape is that we are no longer directing our attention to a singular adversary, but many, all with differing capabilities. Because of the advanced capabilities of Next-gen OPIR, the US remains ready to address the ever-changing threat landscape. As Lauren Thompson (2015), national security contributor to Forbes and Chief Operating Officer at the Lexington Institute, writes, "... for god's sake let's not do any harm to this program, because it really is crucial to America's survival in a world where the number of nuclear-armed nations is growing."

Issues with Cost, Schedule, and Capabilities. While this section could easily have an entirely new article devoted to it, it is important to maintain an encompassing, yet surface-level approach in describing the problems that space-based infrared monitoring faces. Issues stemming from cost, schedule, capabilities, and the need for nuclear-hardened payloads and buses are aspects that have become points of contention for the Next-gen OPIR program. A consistent lack of clarity and a difference in opinions between the USAF, industry, and Capitol Hill have left the program with an estimated price tag in the $2 billion range, only to get it to competitive design review. Due to this difference in opinion, the program has nearly tripled in cost over the last two and a half years. Even though the Next-gen OPIR program has singlehandedly been the cause of many questionable decisions regarding our missile defense capabilities, issues regarding data usage, compatibility of various ground systems, and capability, it is a system that the US would dare not lose. When referring to "data usage," the OPIR program consumes a vast amount of data that is subsequently disregarded. The OPIR program, as noted before, is a highly complex, safeguarded, and classified program, for good reason. While its primary role is to detect and track missile launches by sensing heat signatures, the satellite does much more than that. Unfortunately, these "other capabilities" quickly become classified. Even though there is little to say on the topic regarding Next-gen OPIR's capabilities, we do know that the data gathered from the programs "other capabilities" is extremely difficult to utilize for actionable intelligence, specifically because of the amount that is collected.

Familiarity with the intelligence cycle, T-CPED, tasking, collecting, processing, exploiting, and disseminating helps with understanding the difficulties the Next-gen OPIR program faces in data processing. While the cycle may operate seamlessly, the extreme amount of data has become too much to utilize, subsequently leaving much of it unanalyzed. While both data collection and utilization are a constant challenge, so is the satellite's compatibility with its respective ground systems. As satellites evolve, so must their ground systems, which are responsible for receiving the data captured by the satellite itself. Even though the situation possesses many facets, all the issues above are managed by the USAF and will become one of the first tasks that a separate space branch must address following its establishment. During an interview, Daniel P. Jordan, retired USAF Colonel and former commander of the 2d Space Operations Squadron, states,

We fully support the Air Force's efforts to increase capability, add resiliency, reduce costs, and increase the speed of delivery for our critical national security space assets. We also understand the urgent and increased threats facing our nation in space. Next-Gen OPIR will be an entirely new missile warning system with new payloads and sensors, offering new capabilities to the Air Force or any other newly created space entity.[4]

Weather Observation

The Genesis of Weather Monitoring. Once again, there is a need to return to a tense time in US history, the Cold War. On February 17, 1959, the US Navy's program, Project Vanguard, launched its first weather observation satellite, the Vanguard 2 (Datta 2016). Project Vanguard was responsible for measuring the cloud cover distribution over the daylight portion of its orbit and providing information on atmospheric density throughout its time on orbit. Even though the scientific research gained from creating the Vanguard 2 was widely useful, the data gathered from the satellite was unfortunately unsatisfactory. To compensate for these unfavorable performance characteristics of the Vanguard 2, the Television Infrared Observational Satellite (TRIOS), the "Vanguard 2 follow-on," was launched on April 1, 1960, and is considered to be the first successful weather satellite in history (AFSPC Public Affairs 2018). By the end of 1965, NASA had launched a total of ten TRIOS and gained an approximate 450 useful images. Between 1965 and 1975 programs

4 Derived from a personal interview with Colonel (Ret.) Daniel P. Jordan. When I asked him "what do you perceive to be the largest issue with space-based infrared monitoring, or just the Next-Gen OPIR program in general and does the potential reconfiguration of the current organizational and management structure address this issue?" this is how he responded.

such as the Nimbus, Environmental Science Services Administration Satellite Program, Polar-Orbiting Operational Environmental Satellite, Applications Technology Satellites, and Synchronous Meteorological Satellites contributed to the exploration of satellite weather observation (Datta 2016). On October 16, 1975, the Geostationary Operational Environment Satellite was launched, beginning the lineage of the multi-million-dollar weather observation satellites still in production today (Schmetz and Mezel 2015). While satellite weather capabilities are typically the work of civilian organizations like NASA, the National Environmental Satellite, Data, and Information Service, and NOAA, in 1973, the Defense Meteorological Satellite Program (DMSP), a classified program, was revealed. DMSP provides the Air Force Weather Agency, the intelligence community, and the Navy's Fleet Numerical Meteorology and Oceanography Center with visible, infrared, microwave imagery, temperature, and moisture sounding data, and other specialized space environment data (USAF n.d.).

More than Just a Civil Service. Most importantly, the US needs to collect terrestrial, space environment, and Earth surface data (USAF 2017). Not only is satellite weather data collected and used for providing the civilian population with accurate weather information, but it is also used for protecting both US space assets and their accompanying ground components. Occurrences like thermal flares, radiation emissions, and other potentially harmful phenomena that may negatively affect US space assets are monitored and addressed accordingly because of this capability. They provide intelligence to warfighters and strategic planners alike to ensure that missions may be conducted with little to no "surprise" regarding weather conditions. Space weather assets are responsible for monitoring hurricanes, the polar ice status, vegetation, and oceanic hazards, all of which contribute to the national security and overall well-being of the US.

Supplementing Weather Observation Services. Once again, while there is no dominant singular issue facing space weather, the capability is susceptible to several of the previously mentioned issues. A heavy reliance on commercial supplementation, susceptibility to adversarial intervention, and an array of other possibilities threaten space weather observation. One of the most notable issues has risen from the USAF's questionable ability to provide USCENTCOM with adequate weather data. The issue had risen such concern that the House Armed Services Subcommittee on Strategic Forces directed the Secretary of the Air Force Heather Wilson in the NDAA for FY19 to "develop a plan to provide the United States Central Command with persistent weather imagery for the area of operations of the command beginning not later than January 1, 2026." This directive stems from USCENTCOM's potential reliance on weather data collected from foreign governments. While USCENTCOM has augmented weather data from European nations for over two decades, the systems are aging, and as retired Navy Vice Admiral Conrad

Lautenbacher states, "Using an older satellite to cover the CENTCOM area of responsibility comes with the risk that aging instruments beyond their advertised life spans may fail" (Erwin 2018a). In addition, Ralph Stoffler, the Air Force Director of Weather states, "The challenge in our business is that 95 percent of the data we use comes from the international community... We try to create a balance between what we get from international partners and commercial partners" (Erwin 2018a). Furthermore, a 2017 high-risk report issued by the GAO analyzed a potential weather satellite data gap, concluding that "such a gap could negatively affect military operations that depend on weather data" (GAO 2017). While outsiders typically see weather data as researched-based data with minimal importance to warfighters, they are, in fact, a dire component of the operational ability of US military services.

Space-Situational Awareness and Space Traffic Management

Creating a Catalog. Contrary to popular belief, satellites do not necessarily have to be the manmade objects we commonly think of, but instead, a satellite is the description given to any object in space, natural or synthetic. Before the 1957 launch of Sputnik, individuals like Fred Whipple and G.M. Clemence began researching satellites in space, proving that there was debris in Earth's orbit before humans polluted it. It was not the 1957 launch of Sputnik that initiated space debris, but instead it was the Vanguard 1 and Vanguard 2 that began contributing to the accumulation of space debris (Hall 2014). Because of these launches, the "space object catalog," managed by the Joint Space Operations Center (JSpOC), recently renamed in July 2018 as the Combined Space Operations Center (CSpOC) at Vandenberg AFB in California, was created. CSpOC, and specifically the space object catalog, is a specific USSTRATCOM center responsible for identifying and monitoring all objects in earth orbit (Aerospace 2018). In addition to CSpOC, the Air Force's Eighteenth Space Control Squadron is responsible for operating the Space Surveillance Network, a network that monitors radar and optical sensors at sites located around the world (Aerospace 2018).

The current space situational awareness (SSA) and space traffic management (STM) system, most typically known as the "Space Fence," is a surveillance network comprised of three aspects: radar, a telescope, and a space-based surveillance satellite. Space Fence is radar located in the Marshall Islands on the Kwajalein Atoll and is operated by the USAF (Wener 2018). The system provides CSpOC with a constant stream of data about objects in earth orbit (Mola 2016). As objects pass, the radar reports this information to computers located at CSpOC and can subsequently characterize, catalog, and ultimately monitor the object's trajectory.

Monitoring Potential Mishaps. Outlined in the *Joint Publication 3-14* doctrine, the four functional areas of SSA/STM are Detect/Track/Identification, Characterization, Threat Warning and

Assessment, and Data Integration and Exploitation ("Joint Publication 3-14" 2018). One of the most common debates regarding SSA is launch. To launch satellites into MEO, HEO, and GEO, one must go through LEO, an orbit that is becoming increasingly populated. In general, the ability to launch a satellite into orbit without it colliding into other space debris is known as launch collision avoidance (LCOLA). In addition to LCOLA, SSA capabilities have begun to aid in deorbiting satellites, end-of-life/disposal, reentry, human space flight safety, and adversarial satellite detection (Science Applications International Corporation 2016). SSA's relevance to space operations is much greater than its contributions to satellite monitoring; the capability continues to play a major role in ensuring the United States remains uncontested in the space domain.

The Jurisdictional Challenge. There are two primary debates typically associated with SSA and STM: the strategic implications and potential policy challenges. One of the most prominent strategic discussions tied to SSA and STM is the ability of the US to monitor adversarial space assets as they continue to proliferate. As space technology continues to advance, the ability for US adversaries to not only manipulate data coming to and from the satellite, but also to physically alter US space assets, has become a reality. Not only has the possibility of collisions increased, but so has the potential for intentional kinetic adversarial intervention on US space assets. However, the policy debate associated with SSA and STM is not dissimilar to that of SATCOM. A large majority of the space policy community has advocated for shifting from a predominately DoD-ruled SSA and STM structure, to transitioning responsibility to a civil agency, like the Federal Aviation Administration or the Department of Commerce (Vedda and Hays 2018, 7). This DoD vs. civil discussion tends to surface the inherently governmental aspects of SSA and STM, arguing over whether a civilian agency should be entrusted with the level of responsibility that comes with SSA and STM. One side argues that the increased role that commercial industry plays in space operations demands that the DoD remain intertwined with SSA and STM, and the other side argues that the need for transparency outweighs the necessity of a DoD presence. Whichever way the issue is analyzed, it is uncontested that US national security space assets rely on SSA and STM. While supplementing the DoD's effort in SSA and STM with civilian-based agencies is an option, if this information is to become publicly available, there are in-orbit national security space assets that would be put at risk if this information were to become publicly cataloged. In conclusion, the argument remains that the challenge of supplementing national security for commercial capabilities is a challenge that has yet to be solved (Vedda and Hays 2018, 7).

Space is fundamentally intertwined with US military and defense infrastructures. Maintaining the security of US and allied space assets is not just a military endeavor, but also civilian. Previously examined space-based capabilities are targets for US adversaries. Whether these threats emanate from within the organization itself or are susceptible to external influence, they are present. The next section examines the two countries that pose the largest threat to US national security and commercial space assets: China and Russia. When examining the counter-space capabilities of our adversaries, we must continuously be reminded of what is at stake for the US and its allies.

Defining the Threat

> *Denying US space capabilities is a central tenet of adversary strategies designed to diminish our prestige and raise the risks and costs of intervention in regional affairs.*
>
> —(Ret.) Gen. Robert Kehler (April 2018)

In 1967, the Treaty on Principals Governing the Activities of States in the Exploration and Use of Outer Space, including the Moon and Other Celestial Bodies, most commonly referred to as "The Outer Space Treaty," was signed into law. The space treaty explicitly outlines: "States Parties to the Treaty undertake not to place in orbit around the earth any objects carrying nuclear weapons or any other kinds of weapons of mass destruction, install such weapons on celestial bodies, or station such weapons in outer space in any other manner" (UN 1966). Despite adherence to the treaty following 1967, it has since been perverted, contorted, and outright disobeyed. The adversarial space threat has continued to evolve, and a vast majority of military space leaders involved in the fight have acknowledged this fact. Secretary of the Air Force Heather Wilson (2018) states, "We can no longer view space as a function; it is a warfighting mission. We have been charged with making sure America dominates in space, and that is just what we're doing." In addition, General John "Jay" Raymond, commander of the AFSPC, asserts, "This is one of the most critical times in our national security space history – it will be seen as a strategic inflection point" (AFSPC Public Affairs 2018). And finally, General John J. Hyten, Commander of USSTRATCOM, says, "I watch what our adversaries do. I see them moving quickly into the space domain, they are moving very fast, and I see our country not moving fast, and that causes me concern" (Erwin 2017). While the United States has not yet relinquished its superiority in space, it has fallen behind, and the previously mentioned testimonies justify this fact. This article seeks to examine military doctrine and the space-based capabilities associated with the two countries currently posing the largest threat to United States space superiority, China and Russia. It is important to mention that while Iran, Syria, and North Korea all threaten the US, the largest of these threats comes from China and Russia.

Weaponizing Space

Before analyzing the counterspace capabilities of both China and Russia, we must first assess the technologies that make these nations a threat to the United States. The space technologies currently threatening the US will typically be categorized as "counterspace weapons" and will most commonly be segregated into four distinct categories: kinetic/physical, non-kinetic, electronic, and cyber. In general, a kinetic/physical attack is exactly what it sounds like: the destruction of a satellite through physical contact, typically through the use of an antisatellite (ASAT) weapon. The long-term effects that physical attacks have not only on the satellite, but also its orbit, are extensive, as they can create a debris field within a specific orbit. A non-kinetic weapon, typically a laser, high-powered microwave(s), or EMP may also have physical effects on a satellite. However, non-kinetic weapons do not necessarily need to have physical contact with the satellite (Harrison et al. 2018, 3). When referring to an electronic attack, the typical means of intervention are conducted by jamming and/or spoofing various radio frequencies. As mentioned above, jamming is the action of interrupting of a signal. Jamming can occur on the uplink, i.e., the signal delivered to the satellite, or the downlink, i.e., the signal emitted from the satellite to the ground terminal.

However, jamming and/or spoofing may not always have a permanent effect on the satellite, as the jammer may subsequently be turned off. Finally, and undoubtedly the most complex of counterspace weapons, is an attack by cyber means. Without diving too deep into the technical jargon utilized to explain the cyber domain, it may be best to understand what a cyber-attack could potentially do to a satellite instead of analyzing how an attack is orchestrated. It is best explained within the Space Threat Assessment produced by the Center for Strategic and International Studies: "if an adversary can seize control of a satellite through a cyberattack on the satellite's command and control system, the cyberattack could shut down all communications and permanently damage the satellite by expending its propellant supply or damaging its electronics and sensors" (AFSPC Public Affairs 2018, 5). Overall, the implications of a cyber-attack on a satellite are wide reaching and may differ on a case-by-case basis. In general, these four previously mentioned counterspace weapons are the primary means for executing an attack in space. However, these four types of attacks do not encompass the full scope of counterspace weaponry, but merely scrape the surface of what is possible.

China's Military Doctrine

According to a white paper published by the State Council Information Office of the People's Republic of China (SCIOPRC 2016), China's vision is to "build China into a space power in all respects" with the ability to "effectively and reliably guarantee national security" and to "provide support for the realization of the Chinese dream and the renewal of the Chinese nation." In addition, the phrase "in all respects" is a pertinent aspect of this document. While it may not be immediately apparent what the repercussions of this statement are, this assertion directly affects how the US will view and subsequently respond to China's actions in space. Whether the US seeks to adopt a response that is either offensive or defensive, there is a certain level of action that can and will be taken. To understand how the Chinese space threat became so relevant in such a seemingly short amount of time, it's essential that we take note of the Persian Gulf War.

The US military's performance throughout the Persian Gulf War altered the way the People's Liberation Army (PLA) viewed the relevance of information-enabled weapons and the lethality of precision-strike capabilities (Defense Intelligence Agency [DIA] 2019). It was this war that initiated a change in China's military doctrine, noting that the nation would most likely face a "local war under high-technology conditions" (DIA 2019). Throughout the 1990s, China recognized that a war occurring before 2020 was highly unlikely. It was this strategic vision that provided the nation with a "period of strategic opportunity," or more clearly known as a time for military and economic growth. Throughout this timeframe, it has been president Xi Jinping's goal to reaffirm the PLA's overseas role and to provide substantial military growth beyond traditional PLA capacities (ZD 2017a).

The most recent and relevant developments in Chinese military power occurred in 2015 with President Xi Jinping's structural and organizational reformation of the PLA. This series of military reforms ordered by President Xi Jinping instituted joint theater commands and a new Joint Staff Department and disassembled the previously existing four general departments of the PLA and separated them into fifteen Central Military Commission departments (OSD 2016). This structural reform also elevated China's missile force into a stand-alone service by establishing the PLA Rocket Force, a move that unified China's space and cyber operations under the strategic support force (OSD 2017). In addition, in October 2017 President Xi Jinping outlined China's military goals for the next couple of decades within a report provided the nineteenth Party Congress. The three primary objectives noted by President Xi Jinping were to evolve China into a mechanized force with increased informative and strategic capabilities by 2020, fully modernize the force by 2035, and become a worldwide first-class military power by mid-century (ZD 2017b). While one can speculate as to why China felt it was necessary to elevate its military space operations to such a level through the institution of a "Rocket Force," the answer most likely lies within China's threat perception of its external environment.

In May 2015, the SCIOPRC published China's Military Strategy, a doctrine asserting China's near and long-term military objectives. An alarming aspect of this document is the intense level of concern that China has given to "maintaining peace." Located directly in the preface of this 2015 report is China's prioritization of peace and the seemingly "defensive" military posture it wishes to maintain. However, this report notes "A prosperous and stable world would provide China with opportunities, while China's peaceful development also offers an opportunity for the whole world" (SCIOPRC 2015).

Throughout this report, the importance placed on the word "peace," specifically for China's development, is alarming. However, it is not necessarily the word "peace" that is most alarming, but rather the blatant contradictions outlined within the report, which are both thought provoking and concerning. While the report attempts to argue that a "strong national defense and powerful armed forces" for China offers opportunity for the whole world, it really only seeks to ensure that China's "adaptive" new armed forces "firmly follow the goal of the Communist Party of China (CPC)," not the entire world (SCIOPRC 2015). Even though the report states that it is China's goal to maintain global peace, it also acknowledges that China seeks to "achieve the great rejuvenation of the Chinese nation," regardless of the effects this mission may have on external nations (SCIOPRC 2015, 3). This "rejuvenation" is exemplified through China's digital silk road initiative, its belt road enterprise, and more generally, its increased influence abroad. China's perception of its external environment is categorized within the report as being "generally favorable," despite noting that territorial disputes over the South China Sea have created "grave concerns," and that globalization and revolutionary military affairs have "posed new and severe challenges to China's military security" (SCIOPRC 2015, 4). It seems as though "peace" is the least of China's concerns; instead, security by any means necessary takes precedence.

It would be naïve to discredit the importance and strategic utility of the words within the 2015 report; however, it must also be noted that it is generally the objective of most rational nations to ensure the security and prosperity of their people. Even though this may be the case with China, there is a certain level of skepticism that one must maintain when analyzing their military strategy. It is stated that China will pursue a national defense policy that is defensive and "will never seek hegemony or expansion" " (SCIOPRC 2015, 4). Nevertheless, the same report states that the PLA Air Force seeks to "shift its focus from territorial air defense to both defensive and offensive", and that to "expand and intensify its preparation for military struggle (PMS), China's armed forces must meet the requirement of being capable of fighting and winning" (SCIOPRC 2015, 17). Each one of these quotes, located within the same 2015 report, directly contradicts the assertion that China "will never seek hegemony or expansion." While we must not become too overly analytical of China's defense doctrine, as its unspoken goal is to maintain a certain level of ambiguity and clout, we must also refrain from discrediting the importance of this document. In conclusion, China's military doctrine acknowledges that both "Outer space and cyberspace have become new commanding heights in strategic competition" and that "China will keep abreast of the dynamics of outer space ... and maintain outer space security" (SCIOPRC 2015, 12). As we will see below, China has, in fact, begun to stay abreast of the dynamics of space through its development of advanced space-based capabilities, both offensively and defensively.

China's Counterspace Capabilities

The goal of Chinese space operations is to achieve space superiority by "ensuring one's ability to fully use space while at the same time limiting, weakening, and destroying an adversary's space forces" (Lianju and Liwen 2013). Since the beginning of 2000, China has modernized its space-based command, control, communications, computers, intelligence, surveillance, and reconnaissance capabilities and has increased its overall number of satellites, from a just a few to 181 by mid-2016, second only to the United States (Union of Concerned Scientists 2019). It is Chinese belief that in order to maintain a symmetrical level of power to that of the United States, it must also produce similar forces. It has been stated by Chinese scholars that "Whoever is the strongman of military space will be the ruler of the battlefield; whoever has the advantage of space has the power of initiative; having 'space' support enables victory, lacking 'space' ensures defeat" (Lianju and Liwen 2013, 1). Beginning in 2000, China stated its plans to establish a twenty-four-hour, all-weather remote sensing satellite, along with an operational global satellite navigation system by 2020. Since this declaration, China has launched twenty-two navigational satellites, thirty-four civil, military, commercial communication satellites, and has updated its ground terminal infrastructure (Pollpeter et al. 2017, 8).

Despite these seemingly harmless advances in space, the nation has also enhanced its ability to conduct "space attack and defense operations" (Pollpeter et al. 2017, 8; "Space Threat Assessment"). The most apparent and widely discussed example of these advances occurred in January 2007 when China conducted a successful kinetic ASAT test (Federation of American Scientists [FAS] 2014). The test was performed on an inactive Chinese meteorological satellite located in LEO and confirmed that China can destroy space systems in LEO. Not only did this test prove that China could attack various satellite architectures in LEO, but it also created a great deal of debris (Kelso 2007). Despite the test being a success for Chinese counterspace operations, it directly affected the international community by producing an estimated 3,000 pieces of debris, a repercussion that continues to threaten the ISS and other LEO-based satellites architectures (Weeden 2010). Following the 2007 ASAT test, China launched an ASAT system capable of reaching GEO, an orbit that contains a large number of missile warning, military communications, and ISR satellites (US-China Economic and Security Review Commission 2015, 293). China has also conducted numerous non-debris producing tests in October 2015, December 2016, August 2017, and February 2018 (Gertz 2018).

On top of the previously mentioned ASAT tests, China has also begun experimenting with another increasingly concerning system, the SJ-12 satellite. While the satellite was most likely used to test remote proximity maneuvers near other satellites, jamming, and other counterspace operations, there is speculation that the satellite may have been a preliminary test for a successive 2016 satellite launch, the Aolong-1 space craft. The Aolong-1's publicly acknowledged purpose was to explore options for removing space debris from various orbits, a task that justified a robotic arm being placed on the satellite. Even though there has been a great deal of speculation regarding the mission of the Aolong-1, the technology could more realistically be utilized to damage or disassemble other satellites (Spaceflight 101 2016). In addition to the Aolong-1 were China's remote proximity maneuvers near an older Chinese satellite in 2010, displaying the nation's ability to enter an orbit and subsequently come into close contact with another satellite. This fact, in correlation with the robotic arm placed on the Aolong-1, has caused grave concern.

A key piece to China's non-kinetic counterspace capabilities is its ability to utilize directed energy technologies to "blind or damage sensitive space-based optical sensors, such as those used for remote sensing or missile defense," as outlined in a recent report from the DNI Dan Coats (2018, 13). Not only has the US asserted what it perceives to be the largest space-based threat from China, but also a paper produced by the China Electronic Technology Group Corporation solidifies the assumptions made by the US. The authors of the report state that US space technologies like the AEHF, WGS, and the Global Broadcast Service (GBS) satellite constellations would be susceptible to China's advanced counterspace capabilities (Lin et al. 2012, 20–22; quoted in Chen n.d., 82). While the report was merely an article produced by Chinese academics, the 2014 attack on NOAA's weather systems removed any doubt of China's counterspace capabilities. The attack was initially revealed to the public as "unscheduled maintenance" by Representative Frank Wolfe, former Chairman of the House Appropriations Commerce, Justice, Science Subcommittee, but was subsequently acknowledged as an attack by Chinese hackers. Overall, the 2014 Chinese attack disrupted the flow of NOAA imagery for approximately two days and displayed Chinese counterspace cyber capabilities.

Concluding Remarks on China

Space has inevitably become "a commanding height in international strategic competition" for China (SCIOPRC 2015). The nation "will keep abreast of the dynamics of outer space" and will "maintain outer space security" (Xu 2016). Perhaps the most solidifying aspect of these facts was displayed on December 31, 2015, with the creation of China's Strategic Support Force (SSF). With the creation of this force, China has highlighted the importance of space and information operations. The SSF's creation was predicated on implementing a more streamlined and effective fighting force. It was an attempt to create a force structure that promotes joint operations across the PLA. Ren Xu (2016), China's national defense spokesperson, asserted that the SSF was created to provide "strong strategic, foundational, and sustainment support to carry out the integration of capabilities" with the ability to "optimize the structure of the military forces and improve comprehensive support capabilities."[5] Yin Zhuo, a retired admiral of the Chinese Navy, described the SSF as "an important force in joint operation whose actions will be integrated with the Army, Navy, Air Force, and Rocket Force" (Yao 2016).

It appears that the creation of the SSF was in direct response to an unconsolidated and dysfunctional Chinese military force structure. As different as our nations may be, and as opposite as our strategic and military doctrines are, there is a symmetrical issue happening within US force structure. The creation of the SSF in 2015 sparked concern among US policymakers and defense professionals alike. It should be concerning that even though the SSF was created approximately three years ago, the US is just now beginning to reorganize its national security space components. The previously mentioned testimonies regarding the US falling behind its adversaries in the space domain, specifically China, are not unsubstantiated claims, but are facts supported by the creation of the SSF.

Russia's Military Doctrine

While the Russian and US space relationship is rich in history, we must focus our attention after the fall of the Soviet Union. Despite the intricacies and potential contributions that a historical perspective may provide a better understanding of Russian space history, this analysis is not a historical one, but rather an assessment of the current space environment.

Released in 2010, the Military Doctrine of the Russian Federation solidifies a centuries-old Russian threat perception. A general understanding of how Russia views its external environment can be found at the beginning of Russia's 2010 military doctrine. The document from 2010 promptly begins by listing "the military dangers and military threats to the Russian Federation" (Carnegie Endowment 2010). It does not begin by providing a general assessment of the current geopolitical environment nor does it begin by assessing aspects of Russian military strategy; instead, it commences with an assessment of what it perceives its "main external military dangers" to be. Among these external threats are an expanding NATO presence, an ever-increasing western influence, violations of international treaties, and the militarization of outer space (Carnegie Endowment 2010). Not only does the doctrine immediately address these external threats, it subsequently dives directly into what it's "main military threats are" (Carnegie Endowment 2010). The document notes that a deteriorating military-political situation and a slew of threatening military exercises near Russia's border are "impeding on the operation of systems of state and military command and control" to include "the disruption of the functioning of its strategic nuclear forces, missile early warning systems" and "systems for monitoring outer space" (Carnegie Endowment 2010).

5 Upon finding references pointing to this source, I used Google translate to translate the original Mandarin into English.

Despite being published nearly a decade ago, the 2010 document notes that the characteristics of contemporary military conflicts are evolving. It addresses that an increased reliance on information warfare and proliferated "resources operating in airspace and outer space" are contributing to this evolving threat landscape (Carnegie Endowment 2010). Before the release of the document, on February 12, 2008, both China and Russia submitted a treaty proposal entitled "Treaty on the Prevention of the Placement of Weapons in Outer Space, the Threat or Use of Force Against Outer Space Objects (PPWT)" to the United Nations Conference on Disarmament (Nuclear Threat Initiative 2017). However, the United States dismissed the proposal, dubbing it "a diplomatic ploy by the two nations to gain a military advantage" (Nuclear Threat Initiative 2017). The treaty's denial may have contributed to the aggressive language related to outer space within Russia's 2010 military doctrine. Even though the treaty proposal may have looked like an attempt to make peace, this was not the case, as seen with Chinese and Russian counterspace activities.

Almost symmetrical to China's reorganization activities, just one year later on December 1, 2011, Russia consolidated its air-defense and space forces into a new branch of service known as the Aerospace Defense Forces (ADF) (Ministry of Defense of the Russian Federation n.d.). Four years after this merger, the nation once again reorganized its space organizational structure by combining its Air Force and ADF into the Russian Aerospace Forces (Bodner 2015). The reorganization of these forces was most likely in response to the 2008 PPWT denial and adherence to the direction provided by the 2010 Russian military doctrine. The differences between Chinese and Russian military doctrine is rooted in the two nations' cultural differences. During a recent Center for Strategic International Studies (CSIS) panel featuring Lieutenant General Samuel Greaves, Director of the Missile Defense Agency (MDA), and Dave Trachtenberg, Under Secretary of Defense for Policy (USDP), the two men noted that the major variance between the two nations is Russia's propensity for action. The panelists asserted that despite Russia's weak economy, its inherent inclination to act in offensive and defensive manners is as equally alarming as the advanced Chinese capability (CSIS 2019). Both Lt. Gen. Greaves and Under Secretary Trachtenberg noted that Russia's external threat perception is alarming, concluding that their decision-making calculus is heavily influenced by how the nations perceives its place on the international stage.

Taking into consideration 2010 military doctrine and the reorganization of Russian military space components, the nation may be actively seeking defensive mechanisms for space that err on the side of the offense. Listed below are the capabilities, tests, and instances supporting the reason that Russia's space presence poses a threat to US national security.

Russia's Counterspace Capabilities

Unlike China's ASAT capabilities, Russian kinetic counterspace weapons began in the early 1960s. Between 1963 and 1982, Russia had successfully conducted twenty ASAT tests with its Istrebitel Sputnikov, or "satellite destroyer" (Siddiqi 1997). Throughout the 1980s, 1990s, and early 2000s, Russia experimented with and fielded its most powerful ASAT weapon to-date, known as the Naryad. The system was relocated to Tajikistan in the early 2000s and has since undergone numerous hardware and software updates (FAS 2014). Most recently, the nation has been engaging in what is known as rendezvous and proximity operations (RPO), similar to China's remote proximity maneuvers. A 2018 Secure World Foundation report notes that Russia has conducted various classified RPO activities since 2013, some of which could potentially be threatening (Weeden and Samson 2018). However the most recent of Russia's counter-space weapons may not exactly be classified as "counterspace," but rather "counter-aerospace." The nation has begun to rely on its S-300 and S-400 surface-to-air-missiles and subsequently sold them to its neighboring nations, including Turkey, so that the system could be placed along the Syrian border (Bekdill 2018). Not only has there been a discussion on what will come of the S-400 system, but also the Deputy Commander-in-Chief of the Russian Air Force noted that the S-500, the follow-on system to the S-400, would be made available shortly (Pike n.d.).

In addition to its kinetic counterspace capabilities, Russia has also developed and tested directed-energy and laser weapons systems. Most recently, photos leaked from a 2011 show indicate that the counterspace laser system(s) would be mounted on an aircraft capable of flying at extremely high altitudes and be used to interrupt the functionality of various satellites. In addition to directed energy capabilities, the nation has shown its affinity for GPS jamming technologies, specifically those witnessed during the 2014 annexation of Crimea. Reports emanating from Ukrainian intelligence noted that phones reliant on GPS, radios, and remotely piloted aircraft were all affected as a result of Russia's intervention (Sukhankin 2017). While not directly related to Russian space capabilities, that nation has shown a knack for cyber warfare. Witnessed during the 2014 Crimean conflict, the 2016 US presidential election, and potentially the 2018 midterm elections, Russia has displayed its ability to utilize cyber means as a way of warfare. Even though the attacks were not directly aimed at US space systems, it is safe to say that Russia possesses cyber capabilities that pose a significant threat to the US (Gady 2018).

Concluding Remarks on Russia

As noted earlier, it is not necessarily the capabilities of Russian space assets that are most alarming to US policymakers, but rather the nation's inclination for action and rhetoric surrounding the potential re-weaponization of space through weapon systems like the RS-128, a thermonuclear-armed ICBM with an expected deployable date sometime in 2021 (Gady 2018). Even though Russia is operating within a struggling economic environment, its threat perception of its outside environment displays the nation's dire concern with western influence. Russia has continuously attempted to create distance between itself and the west along its borders. It is this ideology that seeps not only into the sea, land, and air domains, but space as well. If the nation is willing to go to such lengths to maintain a safe distance from western influence along its borders, it is safe to assume that this applies to space as well.

Policy Positions

> *You need to identify, each of you, the key decision makers, the chains of command and empower them to decide quickly.*
>
> —Dr. Michael Griffin (August 8, 2018)

Thus far, we have evaluated the historical framework that has made military space operations possible, we have assessed the reasons that protecting US national security space assets is necessary, and we have analyzed the space capabilities and the potential intentions of both China and Russia. There has been an adequate presentation of the policy issues plaguing specific space capabilities, and we have looked at a few of the challenges regarding the current organizational and management structure of US national security space components. Although specific policy positions and issues will be analyzed, we must begin by assessing the perceptions of the two most powerful players in the space domain, Congress and the DoD. While the US Intelligence Community, specifically the NRO, plays a large role in the current organizational structure of national security space components, the opinions and testimonies coming from the NRO are rarely made available to the public, as their reasons for reorganizing specific national security space components remain classified.

The following section addresses the overall position each organization takes on reorganizing US policy regarding military space operations. While the positions taken by the individuals within each organization differ, we must attempt to evaluate the entities perception in its entirety. One key aspect to consider with Congress is the position of partisanship. This issue of a potential "space force" has become an inherently political development, as the administration mandating this change has been unlike any other. Case in point, Doug Loverro, former Assistant Secretary of Defense for Space Policy, notes, "This has become a partisan issue. I have been saying it shouldn't be, but unfortunately it has become a partisan issue" (Erwin 2018f). Despite the political divide and potential gridlock that this idea will create, it is worth analyzing the positions of Congress and the DoD on the issue through analysis of key documents and testimonies.

The US Congress

To begin analyzing where the direction to assess the current organizational and management structure of the US policy relate to military space operations came from, we must rely on two pertinent documents: the "OMB Report on the Leadership, Management, and Organization of the Department of Defense's Space Activities" (OMB 2017) and the "Final Report on Organizational and Management Structure for the National Security Space Components of the Department of Defense" (DoD 2018). These two reports, while conducted by two separate entities, were presented to the same cohort of individuals that comprise the various congressional defense committees. These committees included the House and Senate Armed Services Committees, the House and Senate Appropriations Committees, their accompanying subcommittees, and the House and Senate Intelligence Committees. It must be noted that these reports may have potentially been the basis for forming the opinions of many key congressional members. Despite a common belief, the pressure for the OMB report originated in the Obama Administration, not the Trump Administration. The OMB report was produced in response to section 1616 of the NDAA for FY17, which would have been largely influenced by the Obama administration. While the Trump administration has had a large influence on assessing the current national security space structure, it is important to note that it was per the guidance of officials from the previous administration. In addition, Congressman Mike Rogers (R-AL) addresses this topic, stating, "People that were not paying attention think the President's Space Force idea came out of nowhere. No. Congress spent the last three years studying this" (Erwin 2018b). A general lack of historical context happens to be a point that is often overlooked.

Section 1616 of the NDAA for FY17 required the director of the OMB to provide recommendations to "Strengthen the leadership, management, and organization of DoD with respect to the national security space activities of the Department ..." The report was produced in response to an already fractured and disaggregated space organizational and management structure. The response, a December 4, 2017 OMB report, asserts that the first recognition of a problematic national security space enterprise was acknowledged by the GAO in 1994 when it asserted that there were "fragmented responsibilities" within the organizational structure (GAO 1994). It also notes that reports conducted after the 1994 GAO report acknowledge that the national security space structure had "Scattered authorities, conflicts of interest, and a lack of consolidated space cadre" (OMB 2017, 4).

However, the 2017 OMB report produced three key findings. This first finding was that DoD space acquisitions management and oversight is fragmented, with many organizations having significant responsibilities (OMB 2017). The second finding notes that the DoD has generally not made significant changes to space leadership over the past two decades (OMB 2017, 7). And lastly, the report concludes that a fragmented leadership structure has contributed to poor coordination and lengthy decision-making (OMB 2017, 7). Given these facts, it is not surprising that the Trump administration has begun to lay the groundwork for the reorganizing the current space management structure, as it would be borderline negligent for the administration not to do so. However, the December 4, 2017 report was not the only piece of analysis provided to Congress that concluded that there were major issues within the management structure of national security space components.

In addition to the 2017 OMB report, section 1601 of the NDAA for FY19, ordered the Deputy Secretary of Defense to "conduct a review and identify a recommended organizational and management structure for the national security space components of the DoD, including the AFSPC, that implements the organizational policy guidance expressed in this section and the amendments made by this session." We must remember that it was acknowledged by both Congress and the executive branch that there were challenges within the current US national security space organizational and management structure, all of which occurred before President Trump's March 13, 2018 speech at Marine Corps Air Station Miramar. It is necessary to surface this fact only because the public dialogue regarding a "Space Force" was largely influenced by this speech, not by the OBM or the DoD reports. The DoD report responding to section 1601 of the NDAA for FY18 was entitled the "Final Report on Organizational and Management Structure for the National Security Space Components of the Department of Defense" and published on August 9, 2018.

The August 9 report provides a four-pronged approach for obtaining the Trump administration's goal of establishing a sixth military branch, a Space Force. Within this document, it is recommended that a Space Development Agency (SDA) be created, followed by a Space Operations Force, then an entity for Services and Support, and lastly, a Space Command. In response to the congressional guidance provided in the NDAA for FY19, the report notes that the DoD "will take immediate steps to implement the President's direction where authorities exist and seek legislation from Congress to realize the President's vision" (DoD 2018). This statement is not merely the DoD arguing on behalf of President Trump's wishes, but rather an attempt to convey the very real threat that our nation is facing in space. It is noted in this report that "Congress has also made its intent and support clear, providing direction and significant funding to enhance national space capabilities" (DoD 2018). Despite the apparent recognition of the threat on behalf of Congress, the idea of a Space Force has been met with opposition throughout its development.

Chairman of the House Armed Services Committee (HASC), Representative Adam Smith (D-WA), asserted that he is "opposed to president Trump's proposal for a Space Force" (Koren 2018). However, Congressman Smith proclaimed that what he opposes "is a separate branch," because he does not "think a separate branch makes sense" (Tritten 2018). Not only have

Congressional members of the Democratic Party opposed the idea, but so have Senate Republicans. Senator James "Jim" Inhofe (R-OK), Chairman of the Senate Armed Services Committee (SASC), stated in December 2018 that "time and time again, ever since this subject came up, I've said there are two things you have to answer. One is, is it going to do a better job than we're doing today? And then two, it's going to cost more—how much more money is it going to cost?" (Kheel 2018). Senator Inhofe went on to say that until he hears answers to those questions, he "will be opposing it, but that doesn't mean it's not going to happen" (Kheel 2018). Not only do the two gentlemen sit on separate sides of the aisle, Inhofe went as far as to say that he and Congressman Smith come from "two different backgrounds," but he thinks that "this is a good example of something that we agree on" (AFSPC n.d.). Despite this unification, the idea has become an inherently political one and is still heavily dependent upon personalities of key policymakers.

While opposition to the idea has been abundant, so too has support. Congressman Mike Rogers, sitting on both the Strategic Forces and the Readiness subcommittees for HASC, has stated that he wants "to get space out of the Air Force bureaucracy and out of a subordinate position" (Erwin 2018b). Despite his advocacy for the idea, Congressman Rogers does note that the DoD will need "to do it responsibly," adding, "we don't need to be too disruptive" (Kheel 2018). In addition, Congressman Jim Cooper (D-TN), Chairman of the Subcommittee on Strategic Forces for HASC, has been a long-time proponent for reorganizing national security space components (Trevithick 2017). Not only have the most recent developments in space reorganization gained congressional support, but in 2017, the idea of a space corps gained bipartisan support from Republican Congressman Mac Thornberry, and his Democratic counterpart, Congressman Adam Smith (Trevithick 2017). While congressman Adam Smith recently opposed the proposed Space Force, he acknowledges that strides must be made to advance the current space organizational and management structure (Trevithick 2017).

Given the various sides and opinions on a new space entity, it's important to remember that this objective was introduced during an inherently political climate. Due to the 2018 midterm elections, a transition of power to the Democratic Party occurred. All the while, the US is still operating under a Republican-led Senate. This fact almost undeniably supports the notion that there will be gridlock when it comes time for Senate approval. It is also important to note that the 2020 election cycle is just around the corner and will be sure to complicate matters even further. In addition, the DoD is currently operating under the Budget Control Act (BCA) of 2011. This act severely hampers the DoD's ability to advocate for an entirely new military branch due to a stringent budget cap. Even though a large majority of the FY20 appropriations, an estimated $170 billion, will be designated under Overseas Contingency Operations, it will be difficult to find the necessary funding necessary to create an entirely new branch of the military.

Congressional views on the topic vary. As mentioned previously, this topic is largely dependent on the individual personalities of our nation's policymakers. While policymakers will never reach complete cooperation on the topic, we must hope that our lawmakers take into consideration the implications of this decision on US national security. Even though congressional support for a "Space Force" seems to be unattainable, congressional support for reorganizing and better structuring our national security space components should be achievable. Also, despite this initiative being an inherently "Trumpian" idea, it is supported with historical examples like the 2001 Rumsfeld Commission and the 2010 NSSS, which each called for some level of action to be taken regarding US policy for military operations in space. In addition, the claim that standing up a Space Force would cost an estimated $13 billion has recently been subdued by the president's budget request for FY20 for approximately $72 million worth of funding necessary to begin this initiative (Whitehouse 2019). So, while both the current administration and the potential cost have each contributed to congressional opposition to the idea, these are issues that our lawmakers will have to address when confronted with the Space Force proposal that will respond to the newly enacted Space Policy Directive 4. Hence, it is not the views or the opinions of individual legislators that matter, but rather what actions the institution will take that mean the most.

Department of Defense (DoD)

While the opinions of the DoD play a key role in reorganizing national security space components, it's important to remember that the department remains subordinate to the commander in chief and will ultimately execute the tasks that it is given. Just as we've seen with Congress, reorganizing US national security space components is not an idea favored by all.

But once again, at the direction of the executive branch, it is not the department's duty to have "opinions" on the topic, but rather to assess how it can ensure a Space Force comes to fruition. Section 1601 of the NDAA for FY18 directed the DoD to assess how it would create a Space Force. The two documents that must be analyzed regarding DoD's perception of a potential Space Force are a memorandum released on September 10, 2018 by Patrick Shanahan and a subsequent memorandum released on September 14, 2018 by Secretary of the Air Force Heather Wilson.

On September 10, 2018, Patrick Shanahan, Deputy Director of the DoD, published a memo entitled "Space Organization and Management Tasks" (Erwin 2018d). The memo acknowledges that section 1601 of the NDAA for FY18 provided direction to the Secretaries of the Military Departments, the Chairman of the Joint Chiefs of Staff, and the OSD for tasks related to space organization and management. Within the memo, various undersecretaries and DoD officials were tasked with addressing the authorities for space, implementing new secretaries for space, producing a proposed timeline for the creation of this branch, and creating potential legislative proposals (Erwin 2018d). Ultimately the report was a demand to key DoD officials to put forth their proposal for a space force. However, not all branches have produced a response to the request like that of the USAF.

In response to Patrick Shanahan's memo, the USAF produced a memo on September 14, 2018: "The Air Force Proposal for a Space Development Agency and Transition to a Department of the Space Force." One of the most prominent quotes derived from this report is located in the second paragraph: "This changing environment affects all capabilities and Military Services. This is a strategic problem we must solve" (USAF 2018). This excerpt explicitly notes that this is an issue of national security. To further advance this point, the first section of the report, "An Approach to a More Lethal Force," rings similar to the 2018 National Defense Strategy (Tossini 2017). This is necessary to note simply because the report doesn't begin with an assessment of what is wrong or what the organizational challenges are, but rather, it outlines exactly why this objective is necessary, to build a more lethal force. The report begins with three immediate recommendations:

1) Assign the Space Rapid Capabilities Office (Space RCO) the function of the SDA, using existing resources and authorities, with the mission of providing space superiority capabilities.
2) Re-integrate defense space and the NRO under the Secretary of Defense's authority, following Senate confirmation.
3) Immediately plan for the resources to establish the Space Force Headquarters in FY2020 (OMB 2017; USAF 2018).

Essentially these objectives address three key facts: an acquisitions cycle tainted by longevity, a disaggregated cadre operating under a disassociated force structure, and the need to devote an appropriate amount of resources to the national security space mission. These recommendations sound eerily similar to the reports above produced by the GAO, which concluded that our space organization and management structure has "scattered authorities, conflicts of interest, and a lack of consolidated space cadre" (Armstrong 2018). While these points may seem monotonous to mention, it is important to understand that these issues are not new developments, but challenges that have inundated the national security space structure for decades.

Despite the documentation and a congressional call to action, many within the department have staunch opinions, and some are opposed to the idea. General John J. Hyten, Commander to USSTRATCOM, noted, "I think that someday we'll have a Space Corps and a Space Force in this country. But I don't think the time is right for that right now" (L.C. Williams 2018). However, as combatant commander, Gen. Hyten may disagree with the Commander in Chief's idea of a potential Space Force. In addition, one of the more polarized answers from Gen. Hyten on the topic was his response to a SASC hearing that occurred in March 2018, noting that he was "not too keen on" the idea of a Space Force, stating once again that this wasn't the appropriate time for this type of reformation (Smith 2018). While top DoD officials currently serving in the department may not have the luxury of overtly displaying their opposition to the idea, former officials do. In September 2018, former Secretary of the Air Force Deborah Lee James warned that creating a Space Force "will sap resources away that could otherwise go to capabilities" (Bender and Kilmas 2018). She further ventured to speak on behalf of current Air Force leaders, asserting that "None of them are in favor of a Space Force – I say none of the top leaders – but they're stuck" (Bender and Kilmas 2018). As mentioned earlier, prompting a top military leader to oppose an idea of the Commander in Chief publicly is a difficult task, as they will typically defer, divert, and outright disregard those types of questions. However, while opposition within the department is difficult to find, it is noteworthy to acknowledge that it does exist.

Despite opposition, many in the department are in fact "keen" on the idea. General David L. Goldfein, Chief of Staff of the USAF, noted in July 2018 that he has:

> got a president of the United States that's talking openly about space as a warfighting domain. I've got a vice president of the United States that stood up a National Space Council and is moving that. I've got Congress that's engaged and now interested in talking a lot about space. I've got the Secretary of Defense working space. I've got a Deputy Secretary. So, I see this as a huge opportunity right now that we've been given to have a national level dialogue about where we're going in space and so I love the fact that the president is leading that discussion. (Seyler 2018)

In addition, Under Secretary of Defense for Research and Engineering Michael Griffin spoke promisingly about reorganizing the current space organizational structure. In response to USD Patrick Shanahan's September 10 memo, Dr. Griffin provided a September 20 report, outlining his proposal for an SDA. Dr. Griffin asserts, "The Space Development Agency is one of the tools we offered up as a way that we're going to reenergize the space development culture, shorten the time cycles that we talked about, bring some new things to the table. That was part of our response back to Congress in the 1601 report" (Erwin 2018e). Once again, Dr. Griffin's response directly addresses the latent acquisitions cycle under which the space industry is currently operating, noting that he wishes to "shorten the time cycles that we talked about" (Erwin 2018e). Whether the current congressional calls to action have created these proponents or they are a result of the individual's honest opinions is unknown. However, it is fair to assume that there are individuals within the department who believe that reorganizing national security space components will make a positive contribution to US national security.

The positions of both Congress and the DoD play a large role in what will ultimately happen with US national security space components. While congressional testimony provided by the DoD may have an impact on the current space organizational structure, this decision is largely reliant upon the choices that members of Congress will make. It is estimated by the USAF that the amount of funding necessary to create a Space Force is approximately $13 billion (USAF 2018). The estimated $13 billion in funding would cover the resources necessary to transfer mission functions, construct a headquarters, realign personnel, and create the necessary installations and facilities (USAF 2018). Taking into consideration the current political climate, the implications of the 2011 BCA, and the scrutiny this idea will face before a Congress that is inadequately versed on space, the reorganization of national security space components will not take place without a struggle. However, Congress must continuously be urged on the seriousness of this issue and be reminded that while it has become politicized, US national security outweighs the political success, or lack thereof, of a single administration.

Policy Positions and Recommendations

> *In my thirty-four-year career in the Air Force, I've never seen such agreement on the importance of space.*
>
> —General John J. Raymond (May 24, 2018)

Despite what a large majority of the public rhetoric conveys, the argument for reorganizing national security space components is more a demand for policy change than a call for creating an entirely new military branch that complicates the already fractured and difficult bureaucratic processes of pre-existing military services. However, if changes to US policy regarding military space operations are going to be implemented, some level of force restructuring will have to occur. And, while sensible administrative reforms are helpful in any endeavor, in the highly technologically complex area of space applications, advancement of R&D, streamlined acquisitions processes, and preparation of a qualified cadre may be directly affected by the following suggested bureaucratic adjustments.

Taking into consideration the threats and challenges mentioned above, there are three specific policy positions that must be addressed to enhance the organizational and management structure of US policy related to military space operations. Not only could these changes create clear lines of authority, but they may also allow for an increased role of both US allies and commercial industry. As a result of these policy changes, the US national security space establishment would be better able to address the advancing adversarial threat. These three recommendations were chosen as a result of the research and analysis provided throughout this article. While not all of the threats and challenges mentioned throughout this article are addressed, the greatest of these challenges are. Throughout this section, when suggestions regarding the reorganization of various space activities are given, no analysis regarding the movement or reassignment of specific individuals within the chain of command is given. An analysis of this type is simply out of the scope of this article. While the following discussion addresses the newly created Major Force Program (MFP) for space, this is largely be assessed as a means for more clearly transitioning space activities from one organization to the next. The following policy recommendations attempt to provide solutions to the major issues mentioned throughout this article.

Recommendation One: Reinstating a US Space Command to Address the Disaggregated Environment

Re-instituting a US Space Command as an eleventh functional unified Combatant Command is the first step in addressing the disaggregated nature of national security space components. Numerous reports have concluded that a non-unified military space community has dampened the United States' ability to remain superior in this domain. Since 2002, after decommissioning USSPACECOM, national security space activities have fallen under the purview of USSTRATCOM. While USSTRATCOM has not directly caused the seemingly diminished US national security space presence, it has not given this domain the attention that USSPACECOM could have provided. During recent congressional testimony, when prompted to speak on the transition of the space mission from USSTRATCOM to the new USSPACECOM, Gen. Hyten highlighted the fact that as commander of USSTRATCOM, space will never be his number one priority (Heyten and O'Shaughnessy 2019). Gen. Hyten asserted the importance of having a new command with a leader that is focused on the military space mission "24 hours a day, 7 days a week" (Heyten and O'Shaughnessy 2019). As a result of this analysis, instituting USSPACECOM is the first step in centralizing national security space activities. As outlined by Secretary Heather Wilson, re-instituting a USSPACECOM will:

- Integrate space planning and operations across military campaigns and contingency plans.
- Simplify the command structure by aligning operational forces to the commander responsible for joint space warfighting.
- Develop space doctrine, concepts of operation and space tactics, techniques and procedures.
- Establish enterprise standards to be adopted by the military services, ensuring interoperability of the joint force.
- Utilize commercial practices and digitization to streamline the foot-

print and automate labor-intensive operations (USAF 2018).

In order to ensure a streamlined implementation of USSPACECOM, both USSOCOM and USCYBERCOM should be assessed for how best to create this command. In addition to USSPACECOM, a holistic approach to national security space activities should be taken.

Defense entities such as the National Geospatial-Intelligence Agency (NGA) and the NRO should also be assessed for areas in which they could create transparency between the intelligence community and the military services. While coordination between the joint chiefs and intelligence community does occur, the potential creation of a Space Force would demand an increased amount of communication and integration. While it appears that this level of communication already occurs, we must recount that merely fifteen years ago, upon the release of the 9/11 Commission Report, it was made public just how unconnected US defense, civil, and intelligence agencies were from one another. Given this fact, it is not unreasonable to assume that a certain level of disassociation still occurs.

However, even though creating an eleventh functional unified Combatant Command may not be the answer to creating a holistic national se-

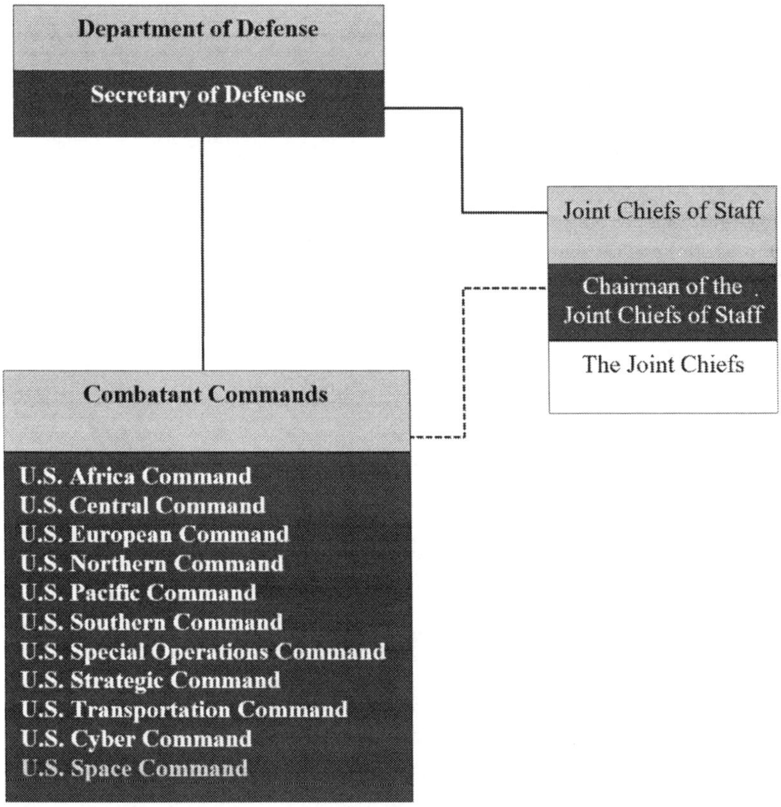

Figure 1. Proposed Addition to the DoD's Combatant Command Structure

curity space environment, it is a step in the right direction. Not only would a combatant command help in the formulation and production of doctrines like the Space Warfighting Construct and the 2011 NSSS, it would also reinstitute a culture rooted in the military space mission (AFSPC). In conclusion, while this policy recommendation has been discussed at length, a December 18, 2018 memorandum instructed the DoD to begin the establishment of USSPACECOM as a functional unified combatant command (Erwin 2019). Figure 1 demonstrates the organizational change that a USSPACECOM would have on the current DoD organizational structure.

Recommendation Two: Creating a Space Development Agency to Provide Rapid Capabilities Development and Expedite Acquisitions through a Whole of Government Approach

The current space acquisitions cycle is a system that typically causes schedule delays and cost overruns. Many within the DoD make note that capability, regulation, and schedule, all of which are largely driven by cost, contribute to a slow space acquisitions cycle. Despite the opinions of many academics, defense policy is not only driven at the requests of combatant commanders and defense doctrines, but is also heavily, if not entirely, influenced by what that year's defense budget is. At the end of the day, the DoD fields what it can, with the resources it has. While the US operates under the world's largest defense budget, it has acquisition and procurement cycles that impede the rapid fielding of necessary capabilities. The challenge to rapidly field capabilities is largely because of a review process that is extremely intensive and trivial. While addressing a room of defense contractors, Dr. Michael Griffin asserted he will ask these companies every chance he gets "to look at what you're doing and find ways to either eliminate it or shortcut it," referring to the acquisitions cycle (Erwin 2018c). During the same forum, Dr. Griffin once again targeted the audience, saying, "You need to identify, each of you, the key decision makers, the chains of command and empower them to decide quickly" (Erwin 2018c).

Using the Air Force's Rapid Capabilities Office and Space RCO as templates for a SDA will set the precedent for instituting this organization. On September 17, 2018, Secretary of the Air Force Heather Wilson asserted, "The Space Rapid Capabilities Office, which was recently established by Congress, provides a mechanism to continue to accelerate special programs of high national priority" (Wilson 2018). To achieve the desired effect, the SDA should move from a heavy dependence on key infrastructure to more proliferated and disaggregated satellite architectures, devote more attention to experimenting with prototypes, and shift from a clustered and overlapping acquisitions structure to a streamlined concentrated structure that generates speed (Secretary of the Air Force 2018). In addition, the US must first begin by altering the rigid culture and archaic practices of its own space community while simultaneously relearning how to build,

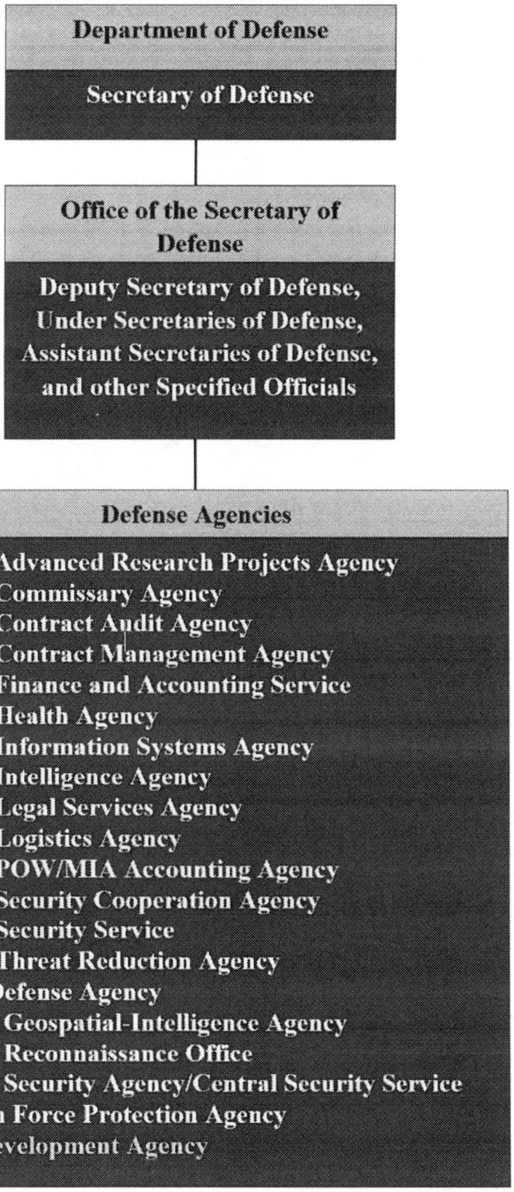

Figure 2. Proposed Addition to the DoD's Defense Agency Structure

deploy, and innovate more rapidly, at lower costs. While dialogue about "lowering costs" has plagued the national security space industry for years, this task is not unattainable. In order to lower costs, the SDA should engage in more Public-Private Partnerships, Cooperative Research and Development Agreements, and other commercial research initiatives. The DoD does not lead industry in groundbreaking technology; instead universities, federally funded research and development centers, and innovative commercial markets do. The security and efficacy attached to this argument is analyzed below.

As previously mentioned, the need to ensure interoperability of satellite architectures is a challenging task. The US currently has architectures that, if afforded the ability to connect, would work seamlessly for various mission sets. However, while the USN may have a component of a space-based architecture, the US Army may have the other and the USAF may have the backbone required to make the systems functional. The funding, time, congressional support, and necessity of these programs is far too great for technology to prevent them from being functional. Creating an SDA that not only centralizes authority, but also ensures that these scenarios do not continue to arise as a demand that must be met. As mentioned in the September 14, 2018 USAF memo, the newly created Space RCO and potential SDA will "consolidate existing efforts within the Air Force to develop key enabling capabilities, including effects, SSA, command and control, and integrations with current operational assets" (Wilson 2018). As General John J. Raymond noted, the Space RCO signaled "a change in capabilities and capacity to get after what we need to do, and that's go fast" (Brissett 2017). The development of an SDA could potentially enhance the Joint Requirements Oversight Council process by addressing the issues within this organization that inadvertently slow down the fielding of new systems (Edwards 2018).

The creation of an SDA, specifically one that is designated as a Combat Support Agency, has the potential to reconfigure a disruptive acquisitions cycle to a more idealistic one, while simultaneously providing opportunities to engage outside entities for lower costs and advanced technologies. The creation of an SDA should take place soon after USSPACECOM is re-established. Placing this agency directly under the OSD as an addition to the already established nineteen defense agencies will ensure its ability to impact the US national security space domain in a positive way. While the exact cost of this agency is not available, the $13 billion worth of funding located in the September 14, 2018 USAF memo accounts for the establishment of this organization (Secretary of the Air Force 2018). Listed below, Figure 2 depicts the organizational change that an SDA would have on the current DoD organizational structure.

Recommendation Three: Utilizing Secure National and Foreign Commercial Services to Expand Influence, Supplement Capabilities, and Enhance Deterrence

Ultimately, this policy recommendation suggests that supplementing current US national security space capabilities with commercial services will better position the national security space establishment to address the advanced threat. While this recommendation does not involve the creation of a new organization, it does stress the importance of USSPACECOM and an SDA. This section focuses on three primary actions: utilizing commercial industry for secure and protected innovative solutions, enhancing international cooperation to strengthen deterrence and expand influence, and most importantly, addressing the advanced space threat.

First, finding ways to better utilize commercial space capabilities for national security purposes is a paramount objective for the US and its allies. It is noted that commercial space capabilities can typically be produced three to five times faster than those of the DoD (Vedda and Hays 2018, 52).

Furthermore, utilizing secure commercial satellite capabilities can augment the costly practice of military services creating, buying, and eventually fielding their own systems. Outlined in this article are the capabilities that would benefit from supplementing secured commercial space-based capabilities. As outlined above, the commercial industry has begun to play a large role in national security space activities and the protection of these systems must be met with an increased level of security. The potential to implement a Civil Reserve Air Fleet (CRAF)-like system for space-based capabilities has been discussed. The CRAF system is a program that selects specific aircraft from US airlines and subsequently uses them to augment DoD airlift requirements in times of emergency (Arnold 2015). While the intricacies of the program would have to be altered for space-based capabilities, it does provide an adequate foundation.

Another goal that must be accomplished to advance national security space capabilities is the need to open, free, and fair competition to new entities and businesses. Streamlining the entrance of new and sometimes smaller players into the space industry, like SpaceX, is necessary. For far too long, the DoD has relied on defense giants that continuously struggle to meet both cost and schedule requirements. There is an array of companies with technologies like disaggregated architectures, smaller satellite buses, and advanced SSA and STM capabilities that would undoubtedly benefit our current national security space posture. The DoD must work to continue fairly competing for these contracts while incentivizing the entrance of new companies to enjoy an industrial and information advantage (Vedda and Hays 2018).

Second, if the US wishes to maintain some level of superiority in the space domain, it must engage its allies. Space incorporated missile defense programs, like the Aegis Ashore, the SM3 Block IIA, and various space-based communications and weather monitoring capabilities, rely heavily upon the cooperation of our allies (C-SPAN 2019). In addition, providing deterrence and ensuring the protection of our international partners from adversarial nations will take the assistance of US allies. Two key components necessary for maintaining a cooperative relationship in space are NATO and the Five Eyes Alliance. The Five Eyes alliance, consisting of Australia, Britain, Canada, New Zealand, and the US, provides coverage of missile tests and foreign satellite deployments and simultaneously monitors the military activities of relevant Air Forces (Tossini 2017). During a 2017 congressional testimony, Lieutenant General David Buck, Commander of the Joint Functional Component - Space for USSTRATCOM, notes that in the operations center of the JSpOC, "we have our allied partners, Five Eyes partners on the OPS floor," asserting that "they are doing to day-to-day, heavy lifting support to the terrestrial fight" (US Congress HASC 2017). Not only will the sustainment of an array of alliances severely affect the decision-making calculus of US adversaries, it will also provide an opportunity to augment certain capabilities like those mentioned above. While many countries that are members of both NATO and the Five Eyes Alliance do not contribute the necessary 2 percent of their gross domestic product, there is an opportunity for the US to supplement the pre-existing space-based capabilities of these nations where necessary (Armstrong 2018). In addition to the United Kingdom, the US can find an ally in Poland, as its President, Donald Tusk, states, "US doesn't have and won't have a better ally than the EU," claiming that "this is an investment in our security, which cannot be said with confidence about Russian and Chinese spending" (Armstrong 2018). While the full cooperation of all NATO and Five Eyes alliance allies will not be attainable to ensure US superiority as it relates to military space operations, there are many US allies who would, in fact, answer this call. In conclusion, deterrence is largely a team sport, and to address the congested, contested, and increasingly competitive environment of space, the US must rely upon its allies that are willing to aid in this fight.

Tying It All Together

While no discussion of a Space Corp or Space Force has been mentioned, this does not mean that an entity of this kind would not positively impact the current national security space enterprise. It is my belief, like Gen. Hyten's, that a Space Corp or Space Force will eventually come to fruition. However, given the analysis provided, an entity of this sort would have to be implemented years down the line, and not in the expedient manner that this administration is proposing. However, the newly created MFP for space would make the transition to a Space Corp or Space Force less disruptive for the national security space enterprise. Given the analysis provided throughout this work, USSPACECOM should be reinstituted and operationally available by the end of FY20.

Subsequently, an SDA should be instated soon after that and before the end of FY21. To end the space acquisitions cycle being consumed by time, an SDA possessing a culture rooted in rapid decision-making authorities, like those mentioned by Lt. Gen. Greaves for the MDA, and an affinity for space superiority would positively benefit our current national security space organizational structure. In addition to these organizations, entities like the Commercial GEOINT Activities (CGA) partnership would aid in these developments. The CGA is a joint venture between the NRO and the NGA, acting as an internet platform for companies to post their space-based and GEOINT-related capabilities that are anticipated to be operationally available within eighteen months of their posting date (CGA 2017). The institutionalization of these types of properly vetted entities would help address and unveil the large amount of talent that is willing to aid in US national security space pursuits.

While the immediate implementation of a Space Corp or Space Force may not be ideal, such developments may be unavoidable and would more than likely have a positive effect in the future. Like the remarks mentioned by Senator Inhofe above, we must ensure that the creation of a Space Corp or Space Force will perform better than the organizational and management structure of US national security space components that is currently in place. And while these answers may not be attainable simply through analysis, the US should begin assessing the potential implications of these developments through the creation of USSPACECOM, an SDA, and supporting organizations.

Lastly, and most importantly, the US must reorganize its national security space components in a way that diminishes the adversarial threat. While there are no simple ways to outright remove the threat, through the previously mentioned policy suggestions, the US can severely diminish adversarial advances by complicating their decision-making calculus and creating an entity that ultimately strengthens the US deterrence posture. In general, deterrence is a game of chess, one that must be played with strategic moves and appropriate actions. The above policy suggestions are both strategically, operationally, and tactically beneficial to the national security space enterprise.

Concluding Remarks

> *Now is the time for our nation to accelerate our efforts to gain and maintain space superiority.*
>
> —Gen. John J. Raymond
> (May 24, 2018)

The superiority of the US national security space presence is being challenged. As seen throughout the historical analysis provided above, the national security space enterprise has been reconfigured, repurposed, and outright neglected. In addition, critical space is not only within the warfighting domain, but also within civil society's most critical capabilities, including infrastructure, energy, information, and finance. And while US space-based capabilities are central to this argument, so are regional threats. It must be reiterated that the paramount objective here is to provide the US with the necessary means and capabilities to address and deter adversaries that continuously attempt to undermine its superiority in space. While this idea may appear to resonate with many, this is not the case. There must be active strides taken by industry, academia, and the private sector to constantly create engagement opportunities and cultivate interest for the national security space enterprise on Capitol Hill. These initiatives must be taken so that the DoD may lessen the number of congressional members unfamiliar with this debate. And lastly, while the threat may never fully be diminished and the US may never have an organizational and management structure that allows national security space components to prosper like other warfighting domains, the aforementioned recommendations would provide the best chance for this to happen.

The first step of reinstating USSPACECOM to address the disaggregated environment and promote a whole of government approach is an action that is already in motion. This recommendation ultimately lays the foundational work necessary to create a culture rooted in the military space mission while centralizing authority over national security space components. Secondly, creating a SDA to provide rapid capabilities development and expedite acquisitions is a mission that must be met so that the DoD can provide seamless interoperability across all branches, for all capabilities. The need to expedite a lengthy acquisitions cycle is not only a call from military service members but is also a demand from those within the top levels of DoD leadership, like Dr. Griffin. In addition, utilizing secure national and foreign commercial services to expand influence, supplement capabilities, and enhance deterrence is a recommendation that inherently leans on US allies. The demand for US superiority in space must be met with help from its international partners, not only to help deter aggression from adversaries, but also to ensure the safety of US allies.

While the recommendations provided are a result of the analysis displayed throughout this work, the policy positions that US Congressional members decide to take will ultimately drive them to decide what happens with the US national security space establishment. While the reports produced by the GAO, OMB, and DoD may have surfaced findings that are unwelcome, they are not unwarranted. What the US Congress decides to do with these facts has yet to be decided. Even though the newly released Space Policy Directive 4 appears to begin laying the foundation for instituting these changes, it is merely a call to action for a construct that Congress will ultimately have to approve. While this administration may never see the establishment of a sixth military branch named the "Space Force," developments toward this idea have come to fruition. However, while a complete reorganization of US policy related to military operations in space may not occur, the national security space establishment must shed the politicized debate that has consumed it while simultaneously receiving the attention and resources it deserves, as this is the real "national emergency" facing this great nation (Baldor 2019).

References

Aerospace. 2018. "Space Debris and Space Traffic Management."

Air Force Space Command. 2013. *Resiliency and Disaggregated Space Architectures White Paper.*

Air Force Space Command. n.d. "Space Warfighting Construct."

Air Force Space Command Public Affairs. 2018. *AFSPC Commander Discusses Strategic Inflection Point for Space at 34th Space Symposium.*

Alver, James G., & Michael P. Gleason. 2018. "A Space Policy Primer: Key Concepts, Issues, and Actors." *The Aerospace Corporation.*

American Institute of Aeronautics and Astronautics. n.d. "History of the GPS Program."

Armstrong, Mark. 2018. "NATO Contributions Country-by-Country." *Euronews*, November 11.

Arnold, David C. 2015. "SpaceCRAF: A Civil Reserve Air Fleet for Space-Based Capabilities." *United States Army War College.*

Balakrishnan, Asha, Becaja M. Caldwell, Reina S. Buenconsejo, & Sara A. Carioscia. 2018. "Global Trends in Space Situational Awareness (SSA) and Space Traffic Management (STM)." *Science & Technology Policy Institute.*

Baldor, Lolita C. 2019. "US Northern Command Leader Says No Military Threat on Southern Border." *Military Times*, February 26.

Barock, Richard T. 1995. "Space Operations and Tactical Application - US Navy." *Space Tracks* winter.

Bekdill, Burak Ege. 2018. "Turkey Defense Minister Announces Timeline for S-400 Deployment." *Defense News*, October 26.

Bender, Bryan, & Jacqueline Kilmas. 2018. "Trump's Space Force Struggling to Launch." *Politico*, September 17.

Bennett, Michael. 2012. "Options for Modernizing Military Weather Satellites." *Congressional Budget Office.*

Berkowitz, Bruce. 2011. "The National Reconnaissance Office at 50 Years: A Brief History." *Center for the Study of National Reconnaissance.*

Berlocher, Greg. 2008. "Military Continues to Influence Commercial Operators." *Via Satellite*, September.

Bodner, Matthew. 2015. "Russia Merges AF with Missile Defense, Space Commands." *Defense News,* August 8.

Boehm, Joshua. n.d. "A History of United States National Security Space Management and Organization." *Federation of American Scientists.*

Boyd, Andrew H. 2017. "Satellite and Ground Communication Systems: Space and Electronic Warfare Threats to the United States Army." November 7.

Bradburn, David D. n.d. "Evolution of Military Space Systems."

Bradburn, David D., John O. Copley, & Raymond B. Potts. 2016. "The National Reconnaissance Office (NRO) History: The SIGINT Satellite Story." *National Reconnaissance Office.*

Brissett, Wilson. 2017. "The Creation of a Space Rapid Capabilities Office." *Air Force Magazine*, December 8.

Buesnel, Guy, & Mark Holbrow. 2017. "GNSS Threats, Attacks and Simulations." *Spirent*, June.

Bullard, John W. 1963. "History of the Field Army Ballistic Missile Defense System 1959–1962."

Carnegie Endowment. 2010. *President of the Russian Federation: 2010 Military Doctrine of the Russian Federation.*

Center for Strategic and International Studies. 2019. *The 2019 Missile Defense Review: What's Next?*

Central Intelligence Agency. 2008. *A Look Back ... The National Security Act of 1947.*

Chen, David D. n.d. *Opening Statement of Mr. David Chen.*

CGA. 2018. *Commercial GEOINT Activity (CGA): Leaderboard 1.1 User Manual.*

Coats, Daniel R. 2018. "Statement for the Record: Worldwide Threat Assessment of the US Intelligence Community." *Department of National Intelligence.*

Cobham SATCOM. n.d. *Satellite Bandwidth 101.*

C-SPAN. 2019. *Pentagon Officials Discussed the 2019 Missile Defense Review, with Recommendations on Space-Based Operations, Hypersonic Missiles, and the F-35 Fighter.*

Cutshaw, Jason B. 2017. "SMDC Celebrates 60 Years of Defensing the Nation." *US Army.*

Datta, Anusuy. 2016. "A Brief History of Weather Satellites." *Geospatial World*, November 19.

DCI Task Force on The National Reconnaissance Office ... (n.d.). Retrieved from https://nsarchive2.gwu.edu/NSAEBB/NSAEBB35/docs/doc14.pdf

Defense Advance Research Projects Agency. n.d. *Transit Satellite: Space-Based Navigation.*

Defense Intelligence Agency. 2019. *China Military Power: Modernizing a Force to Fight and Win.*

Delpech, Theresa. 2012. "Nuclear Deterrence in the 21st Century: Lessons from the Cold War for a New Era of Strategic Piracy." *RAND Corporation.*

Department of Defense. 2018a. *Final Report on Organizational and Management Structure for the National Security Space Components of the Department of Defense.*

Department of Defense. 2018b. *Interim Report on Organizational and Management Structure for the National Security Space Components of the Department of Defense.*

Department of Defense Directive 5160. 32. 1970. *Development of Space Systems.*

Department of Defense Directive 5030. 18. n.d. *DoD Support of National Aeronautics and Space Administration.*

Department of National Intelligence. 1947. *Ref Book-1947 National Security Act.*

Dev, Rishabh. 2017. "LEO, MEO & GEO Satellite Systems: A Comparison." *Durofy*, January 27.

Edwards, Jane. 2018. "Gen. John Hyten: DoD Leadership Understands the Need for Faster Space Procurement Process." *ExecutiveGov*, March 22.

Erwin, Sandra. 2017. "STRATCOM Chief Hyten: 'I Will Not Support Buying Big Satellites That Make Juicy Targets.'" *Spacenews*, November 19.

———. 2018a. "New Concerns About US Central Command's Access to Weather Satellite Data." *Spacenews*, April 26.

———. 2018b. "Rep. Mike Rogers: Space Force Will be Done 'Responsibly' with Minimal Disruption," *Spacenews*, June 21.

———. 2018c. "Mike Griffin's Tough Talk to Pentagon Contractors: Be a Team Player, Look at Your Own Red Tape." *Spacenews*, August 12.

———. 2018d. "New Pentagon Memo Lays Out Action Plan to Establish Space Force by 2020." *Exonews*, September 20.

———. 2018e. "Griffin: Future of new DoD Space Agency 'Still up in the Air.'" *Spacenews*, November 13.

———. 2018f. "Political Odds Stacked Against Space Force. Proponents Brace for Long Fight." *Spacenews*, November 14.

———. 2019. "Shanahan Gearing up for Space Debate on Capitol Hill." *Spacenews*, January 29.

"Fact Sheet: DoD Strategy for Deterrence in Space." n.d.

"Fact Sheet: National Security Space Strategy." n.d.

Federation of American Scientists. 2014. *Sourcebook on the Okno, Okno-S, Krona and Krona-N Space Surveillance Sites.*

Federation of American Scientists. n.d. *Organizations that Affect National Security Space.*

Fogleman, Ronald R. n.d. *The Air Force and the Military Space Program.*

Forest, Benjamin D. 2008. "An Analysis of Military Use of Commercial Satellite Communications." *Naval Post Graduate School*, September.

Gady, Franz-Stefan. 2018. "Russia to Test Fire RS-28 Sarmat ICBM in Early 2019." *The Diplomat*, October 3.

General Accounting Office. 1994. *National Space Issues: Observations on Defense Space Programs and Activities.*

Gertz, Bill. 2018. "China ASAT Test Part of Growing Space War Threat: DNI Outlines Growing Danger to Satellites from Beijing's Missiles, Lasers and Robot Spacecraft." *Free Beacon*, February 23.

Governmental Accountability Office. 2017. *High Risk Series: Progress on*

Many High-Risk Areas, While Substantial Efforts Needed on Others. Report No. GAO-17-317, February.

Grant, Dustin L., & Matthew J. Neil. 2018. *The Case for Space: A Legislative Framework for an Independent United States Space Force.* April.

Guier, William H., & George C. Weiffenbach. 1998. "Genesis of Satellite Navigation." *Johns Hopkins APL Technical Digest* 19 (1).

Hall, Cargill, & R. Jacob Neufeld. 1998. "The US Air Force in Space 1945 to the Twenty-first Century," *USAF History and Museums Program.*

Hall, Loretta. 2014. "The History of Space Debris." *Embry-Riddle Aeronautical University*, November 6.

Harrison, Todd, Kaitlyn Johnson, & Thomas G. Roberts. 2018. "Space Threat Assessment 2018." *Center for Strategic and International Studies,* April, 3.

Hart, Benjamin. 2018. "Trump Announces 'Space Force' He Wants to be Sixth Branch of Military." *New York Magazine*, June 18.

Hartinger, James V. 2011. "High Frontier: The Journal for Space and Cyber Space Professionals." *US Air Force* 7 (4).

Howell, Elizabeth. 2018. "NAVSTAR: GPS Satellite Network." *Space,* April 26.

Hyten, John E., & Terrence J. O'Shaughnessy. 2019. "Senate Armed Services Committee Hearing: United States Strategic Command and United States Northern Command." *Senate Armed Services Committee*, February 26.

Insinna, Valerie, & Aaron Mehta. 2018. "Trump Orders Creation of Independent Space Force – But Congress will have its Say." *Defense News,* June 18.

Jianing, Yao. 2016. "Expert: The Strategic Support Force will be Critical for Victory During the Entire Operation." *China Military Online,* January 6.

Jin-shun, Lin, Wu Xianzhong, Lu Shengjun, & Jiang Chunshan. 2012. "Countermeasure Technology for MMW Satellite Links." *Aerospace Electronic Warfare,* October.

"Joint Publication 3-14: Space Operations." 2018.

Kelso, Thomas Sean 'T.S.' 2007. "Analysis of the 2007 Chinese ASAT Test and the Impact of Its Debris on the Space Environment." *Center for Space Standards & Innovation.*

Kennedy, Gary, & Crawford, Michael J. 1998. "Innovations Derived from the Transit Program." *Johns Hopkins APL Technical Digest* 19 (1).

Kheel, Rebecca. 2018. "Senate Armed Services Chair not Convinced of Need for Trump Space Force." *The Hill*, December 13.

Kim, Yool, Elliot Axelband, Abby Doll, Mel Eisman, Myron Hura, Edward G. Keating, Martin C. Libicki et al. 2015. "Acquisition of Space Systems: Past Problems and Future Challenges Volume 7." *RAND Corporation*.

Koren, Marina. 2018. "Trump's Space Force Faces an Uncertain Fate." *The Atlantic*, November 9.

Kusiolek, Richard. 2010. "Peace of Operations Increases Demand on Satcom on the Move." *Satellite*, April.

Lang, Sharon Watkins. 2015. "SMDC History: ARGMA Opens with a Blast." *US Army*, October 15.

Laurie, Clayton D. 2001. "Congress and the National Reconnaissance Office." *National Reconnaissance Office*, June.

Lee, Ricky J., & Sara L. Steele. 2014. "Military Use of Satellite Communications, Remote Sensing, and Global Positioning Systems in the War on Terror." *Journal of Air Law and Commerce* 79 (1): Article 2.

Lianju, Jian, & Wang Liwen, eds. 2013. *Textbook for the Study of Space Operations*. Beijing: Military Science Publishing House.

Loverro, Douglas L. 2018. *Statement Before the House Armed Services Committee: Space Warfighting Readiness: Policies, Authorities, and Capabilities*. March 14.

Mai, Thuy. 2012. "Global Positioning System History." *NASA*, October 27.

Ministry of Defense of the Russian Federation. n.d. *History: Aerospace Defense Forces*.

Mitchell, John Edward. 1991. "Apogee, Perigee, Recovery: Chronology of Army Exploitation of Space." *RAND Corporation*.

Mola, Roger. 2016. "How Things Work: Space Fence: The New Early-Warning System to Protect Spacecraft from Orbiting Junk." *Air and Space Magazine*, February.

Mosher, Dave. 2018. "Astronaut Mark Kelly says Trump's Plan to Create a Space Force is a Dumb Idea.'" *Business Insider*, August 10.

National Aeronautics and Space Administration. n.d. *Challenges of Military Satellite Communications*.

National Oceanic and Atmospheric Administration. n.d. *NOAA's Geostationary and Polar-Orbiting Weather Satellites*.

National Research Council. 2005. *Navy's Needs in Space for Providing Future Capabilities*.

"National Security Space Strategy: Unclassified Summary." 2011.

"Naval Satellite Operations Center (NAVSOC)." 1997.

"Naval Space Command (NAVSPACECOM)." n.d.

Nuclear Threat Initiative. 2017. *Proposed Prevention of an Arms Race in*

Space (PAROS) Treaty. Last updated September 29.

Office of Management and Budget. 2017. *OMB Report on the Leadership, Management, and Organization of the Department of Defense's Space Activities.*

Office of the Secretary of Defense. 2016. *Annual Report to Congress; Military and Security Developments Involving the People's Republic of China 2016.*

Office of the Secretary of Defense. 2017. *Annual Report to Congress; Military and Security Developments Involving the People's Republic of China 2017.*

PBS. n.d. "A Chronicle of Missile Defense, from the Dawn of the Missile Age During World War II to the Present."

Pike, John. n.d. "S-500 Samoderzhets." *Global Security.*Pollpeter, Kevin L., Michael S. Chase, & Eric Heginbotham. 2017. "The Creation of the PLA Strategic Support Force and Its Implications for Chinese Military Space Operations." *RAND Corporation.*

Resilient Navigation and Timing Foundation. 2016. *Prioritizing Dangers to the United States from Threats to GPS: Ranking Risks and Proposed Mitigations White Paper.*

Report of the Commission to Assess United States National Security, Space Management and Organization. 2001.

Richelson, Jeffery T. 2007. "Space-Based Early Warning: From MIDAS to DSP to SBIRS." *The National Security Archive,* November 9.

Roeder, Tom. 2018. *Space Force: A Timeline.*

Russia Tests an Intercontinental Ballistic Missile. n.d.

Satellite Communications for the Warfighter MILSATCOM Handbook Volume 1. Global Security.

"Satellite Orbits." n.d. *Just.edu.jo.*

Schmetz, Johannes, & W. Paul Mezel. 2015. "A Look at the Evolution of Meteorological Satellites: Advancing Capabilities and Meeting User Requirements." *American Metereology Society,* July 1.

Schulte, Gregory. 2012. "Protecting Global Security in Space," Presentation given at the S. Rajaratnam School of International Studies Nanyang Technological University, Singapore, May 9.

Science Applications International Corporation. 2016. *Orbital Traffic Management Study, Appendix D, D-8.*

Seyler, Matt. 2018. "Air Force Chief of Staff Talks Space Force: 'I Love the Fact That the President is Leading That Discussion.'" *ABC News,* July 18.

Siddiqi, Asif A. 1997. "The Soviet Co-Orbital Anti-Satellite System: A Synopsis." *JBIS - Journal of the British Interplanetary Society* 50: 225–40.

Smith, Marcia. 2018. "Hyten Not Ready

to Endorse Space Force." *Space Policy Online*, March 20.

Spaceflight 101. 2016. *China's New Orbital Debris Clean-up Satellite Raises Space Militarization Concerns.*

Spires, David N., & Rick W. Sturdevant. n.d. "From Advent to Milstar: The US Air Force and the Challenges of Military Satellite Communications." *National Aeronautics and Space Administration.*

State Council Information of the People's Republic of China. 2015. *China's Military Strategy.*

State Council Information Office of the People's Republic of China. 2016. *China's Space Activities in 2016.*

Stone, Christopher Michael. 2015. "Reversing the TAO: A Framework for Credible Space Deterrence." *Missouri State University.*

Sturm, Thomas A. 1967. "The USAF Scientific Advisory Board: Its First Twenty Years 1944–1964." *Historical Division Liaison Office.*

Sukhankin, Sergey. 2017. "Russian Electronic Warfare in Ukraine: Between Real and Imaginable." *Real Clear Defense*, May 26.

Symmetry Electronics. 2015. *What is the Difference Between GNSS and GPS.*

Thompson, Loren. 2015. "SBIRS: The Pentagon's Most Important Space Program for Preventing Nuclear War." *Forbes*, June 8.

Tossini, J. Vito. 2017. "The Five Eyes – The Intelligence Alliance of the Anglosphere." *UK Defense Journal*, November 14.

Trevithick, Joseph. 2017. "A Primer on The Raging Battle for A New Pentagon Space Corps." *The Drive*, July 12.

Tritten, Travis J. 2018. "Rep. Adam Smith Says he Opposes Space Force." *Washington Examiner*, September 13.

Uchill, Joe. 2016. "Why GPS is More Vulnerable Than Ever: The Space-Based Navigation and Timing System Faces a Growing Risk of Attack. But There is a Simple Solution." *Christian Science Monitor*, January 8.

Union of Concerned Scientists. 2019. "UCS Satellite Database: In-Depth Details on the 1,459 Satellites Currently Orbiting Earth." Last revised January 9, 2019.

United Nations. 1945. *United Nations Charter.*

United Nations. 1966. *Treaty on Principles Governing the Activities of States in the Explorations and Use of Outer Space, Including the Moon and Other Celestial Bodies.*

University of Tasmania. 2014. "How GPS Works." Retrieved from http://dpipwe.tas.gov.au/Documents/Worksheet_1_-_How_GPS_Works.pdf

US Air Force. 2018. *The Air Force Proposal for a Space Development Agency and Transition to a Department of the Space Force.*

US Air Force. 2017. *Defense Meteorological Satellite Program.*

US Air Force. 2015. *Defense Satellite Communications System.*

US Air Force. n.d. *USAF Meteorological and Space Environmental Services.*

US Archives. n.d. *Department of Defense. Department of the Navy. Naval Space Command. Navy Astronautics Group.*

US Army. n.d.a. *AMCOM History.*

US Army. n.d.b. *US Army Space and Missile Defense Command/Army Forces Strategic Command: The Army Service Component to the US Strategic Command.*

US Army. n.d.c. US Army Space and Missile Defense Command/Army Forces Strategic Command: Mission.

US China Economic and Security Review Commission. 2015. *2015 Report to Congress of the US-China Economic and Security Review Commission.* November.

US Congress. House Armed Services Committee. 2017. *House Armed Services Committee Hearing on Fiscal 2019 Priorities.* Retrieved from https://armedservices.house.gov/

US Congress. House of Representatives. n.d. *Joint Explanatory Statement of the Committee of Conference for FY19.* Retrieved from https://appropriations.house.gov/

Van Inwegen III, Earl S. n.d. "The Air Force Develops an Operational Organization for Space." In *the US Air Force in Space: 1945 to the Twenty-First Century.*

Vedda, James A., & Peter L. Hays. 2018. "Major Policy Issues in Evolving Global Space Operations." *The Mitchell Institute for Aerospace Studies/Air Force Association*, February.

Wade, Mark. *Nike Zeus: Part of Spartan ABM Family.* Washington, DC: The National Academies Press.

Watkins, Sharon. 2015. "SMDC History: ARGMA Opens with a Blast." *US Army*, October 15.

Watkins, Sharon. 2018. "SMDC History: A-B-M-D-A." *US Army*, March 15.

Natural Resources Canada. n.d. "Weather Satellites/Sensors."

Weeden, Brian. 2010. *2007 Chinese Anti-Satellite Test Fact Sheet.*

Weeden, Brian, & Victoria Samson. 2018. "Global Counterspace Capabilities: An Open Source Assessment." *Secure World Foundation*, April.

Wener, Debra. 2018. "Lockheed Martin Prepares to Turn on US Air Force Space Fence on Kwajalein Atoll." *Space News*, May 3.

Whitehouse. 2019. *A Budget for a Better America: Promises Kept. Taxpayers First.* March 11.

Williams, Lauren C. 2018. "STRATCOM Leader Pushes Back on Space Force Idea." *Defense Systems*, March 21.

Williams, Matt. 2017. "What is Low Earth Orbit." *University Today*, January 6.

Wilson, Heather. 2018. "The Air Force We Need." Paper given at the *Air Force Association Conference*, September 17.

Xu, Ren. 2016. *Ministry of National Defense Spokesperson Takes Media Inquiries on Deepening National Defense and Military Reform.* January 1.

Yiannopoulos, Philip. 2018. "Inside the Epic Debate on Rethinking Our 50-Year-Old Outer Space Treaty." *Fast Company*, September 24.

ZD. 2017a. "China Focus: 'Be Ready to Win Wars,' China's Xi Orders reshaped PLA." *Xinhua* in English, August 1.

ZD. 2017b. "Full Text of Xi Jinping's Report at 19th CPC National Congress." *Xinhua*, November 3.

Commentary

Why Students Should Seek an Internship with SpaceX Amid the COVID-19 Pandemic and What to Expect

Melissa Layne
American Public University System

Introduction

COVID-19 has left global economies struggling as they continue to enact strict measures to slow the pandemic from spreading. Despite many industries making an abrupt but successful transition from face-to-face to virtual, home-based work environments, many others have not been as fortunate and are barely keeping their businesses afloat. There is one industry, however, that is doing an exceptional job of weathering the economic turmoil—the space industry.

Even pre-COVID-19, the space industry had a strong track-record of attracting interest from the investment community. Investment banks including Morgan Stanley, Goldman Sachs and Bank of America Merrill Lynch expect the space industry to grow to over $1 trillion by 2040 (Palerm, 2019). However, in their 2019 report Global Satellite Capacity Supply and Demand, 16th Edition, Northern Sky Research (NSR) predicts that the resulting revenues will be modest. They report that when combining Satcom capacity, satellite manufacturing, launch and ground equipment, the $1 trillion projection is highly achievable, but perhaps may take longer (NSR, 2019), yet perhaps not.

COVID-19 and the Immediate Need for Broadband

Life has changed drastically since the NSR's June 2019 report. The Coronavirus pandemic and mandatory self-isolation restrictions to control its spread have triggered the overwhelming demand for broadband Internet.

There are a few low-earth orbit satellite providers such as Amazon's Project Kuiper and Telesat's LEO satellite constellation, however, they are all several years behind in even possibly offering customers high-speed Internet service. Another provider, OneWeb, recently filed for bankruptcy. Leading in both production and satellite launches is Elon Musk's Starlink (Mack, 2020), which to date has launched 240 satellites and expects to offer Internet globally sometime in 2021 (Cooke, 2020). SpaceX's lead manufacturing engineer for Starlink, Jessica Anderson, states

"The design goal for Starship is three flights per day on average [per ship], which equates to roughly 1,000 flights per year at greater than 100 tons per flight. This means every 10 ships would yield 1 megaton per year to orbit. This is a significant effort, and we are looking for highly skilled engineers and welders to help us make this a reality." (Mazumdar, 2020)

SpaceX's success has strategically positioned them as a leader in the aerospace and satellite industry. Despite COVID-19, SpaceX has been spared from being forced to cut salaries or lay off employees. Conversely, SpaceX continues to hire. Additionally, this success will likely to continue should the dependency upon their ships and satellites continue.

The Need for Space Professionals

Overall, the space industry will continue to fill increasing job positions in both the private and military sectors and diverse educational and training opportunities for the next-generation of space professionals will also be required. Table 1 presents education, experience, and training typically required for space-related occupations (Angeles and Vilorio, 2016).

Table 1. Education, experience, and training typically required for space

Occupation	Typically needed to enter the occupation		Typically needed to attain competency in the occupation
	Education	Work experience in a related occupation	On-the-job training
Scientists			
Astronomers	Doctoral or professional degree	None	None
Atmospheric and space scientists	Bachelor's degree	None	None
Physicists	Doctoral or professional degree	None	None
Engineers			
Aerospace engineers	Bachelor's degree	None	None
Computer hardware engineers	Bachelor's degree	None	None
Electronics engineers(1)	Bachelor's degree	None	None
Mechanical engineers	Bachelor's degree	None	None
Technicians			
Aerospace engineering and operations technicians	Associate's degree	None	None
Avionics technicians	Associate's degree	None	None
Life, physical, and social science technicians, all other(2)	Associate's degree	None	None
Media and communications			
Photographers	High school diploma or equivalent	None	Long-term on-the-job training
Producers and directors	Bachelor's degree	Less than 5 years	None
Public relations specialists	Bachelor's degree	None	None
Technical writers	Bachelor's degree	Less than 5 years	Short-term on-the-job training

Footnotes:
(1) Excludes computer engineers.
(2) Includes includes meteorological aides.
Source: U.S. Bureau of Labor Statistics

In addition to a space studies program, gaining on-the-job experience with an internship is one of the best ways to jump-start a new career.

Internships

Because there are a number of career pathways in the space industry (as shown in Table 1), there exist several internship opportunities with organizations, government, and private companies. Internships are a wonderful way to gain experience, learn from mentors, connect with other interns, and hone valuable skills that provide an added advantage to a student close to graduating, or a recent post-grad. In addition to continually hire, SpaceX also offers an internship to graduate students and recent post-grads who are eager to embark on a space career.

The SPACE-X Internship Program

The SpaceX Internship is a year-round program offered in the Spring (Jan – Apr), Summer (May–Aug), and Fall (Sept–Dec) with opportunities to play a direct role in transforming space exploration and helping realize the next evolution of humanity as a multi-planetary species.

Figure 1. SpaceX headquarters in Hawthorne, California

Opportunities are available across all engineering functions and business operations at various locations including:

- McGregor, TX – Rocket development facility
- Cape Canaveral, FL – Launch site
- Vandenberg AFB, CA – Launch site
- Hawthorne, CA – Headquarters

- Seattle, WA – Satellite development
- Washington, DC – Government affairs

They do not exclusively recruit from certain schools, but rather, they seek the most talented candidates for their advanced Intern Program and entry-level full time positions that have a history of significant contributions to hands-on extracurricular projects, in addition to a strong academic record. Technical interns and associate engineers are involved in student-driven engineering groups in school, that focus on designing, building, testing, and/or coding. For interns in non-technical roles, they look for similar hands-on project experience in their respective fields. Internships and associate engineer opportunities at SpaceX are paid, and can also be used to satisfy course credit at the discretion of the student's university.

Take it From a SpaceX Intern—Here's What They Have to Say

For both internship and associate engineer positions, students learn valuable hands-on experiences. The most successful candidates for SpaceX's Intern Program have a history of significant contributions to hands-on extracurricular projects in addition to a strong academic record. This is not an ordinary "internship"—students are given as much responsibility as regular full-time engineers and are an important part of the team.

Here are some interns, past and present, who share their thoughts on working at SpaceX.

Julie Wang: *"I'm an intern with avionics right now I'm working on a receiver for the Dragon and in order to navigate dragon communicates with Tigress which is a sub-satellite that NASA sent up. Specifically, I'm working on the receiver that goes on Dragon that will link up to those satellites [to] figure out where the capsules will fit and then navigate."*

David Larson: *"I'm in the structure department here at SpaceX. Spacex is pretty awesome. It's an incredible place to work. You get to do things that you don't get to do anywhere else, and as an intern you get to work on it. You don't have to be a full-time engineer to work on technologies and projects that are pushing the boundaries in Space Flight."*

Joshua Mann: *"I work in propulsion. I have programming computer tasks which pretty much is how we determine whether or not our engines are operating to the right specifications. I'll do a lot of analysis on the different devices that you'll use to test them and make sure that they're running in the correct fashion."*

Brendan Hardy: *"I work with ground support equipment engineering in McGregor, Texas. We're really more hands-on get-it-done sort of team. I've learned a ton—especially what we do in ground support equipment. I could take what I've learned and I could go to an oil field, I could go to another aerospace industry. I could go pretty much anywhere with all the stuff that I've learned."*

Jeff Ellenoff: *"I'm a test engineer focusing on Dragon development. I get to come to work and play with spaceships day. What person wouldn't want to do that? When an engine test happens it's the most incredible thing you could possibly think of."*

Katina Mattingly: *"I'm a launch intern at Vandenberg, CA. I work with all the electrical GSE and it's basically command and control of the site so it's a lot of sensor reading and valve control. I have learned a ton about actual hands-on things that I didn't get from my schooling and also working with technicians. I work with technicians every day. My view everyday driving into work is the ocean ... and it's a beautiful view."*

Tejas Parekh: *"I am a launch intern at Cape Canaveral. Walking the launch pad is crazy! There are so many different aspects to it and you get to do absolutely everything here. So I get to fulfill roles as a design engineer working at my desk with modeling and analysis, and then the next day I work as a build engineer ordering materials and sending drawings to technicians. I'd recommend this location to anybody—especially due to the fact that you get to act as an operations engineer and actually launch a rocket!"*

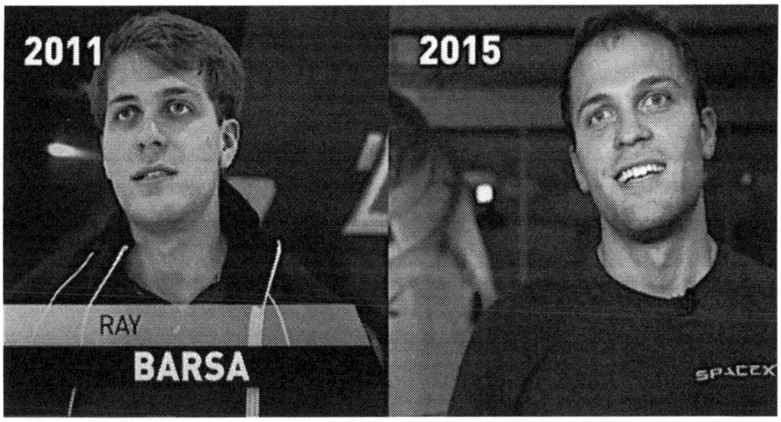

Ray Barsa (former Intern): *"SpaceX is a great place for interns to find full-time jobs. By the end of my internship, I knew all of these processes and approaches that SpaceX took to manufacturing, so by the end of it, I could really kind of take whatever composite assembly was thrown at me and bring it to life here in the factory."*

Josh Green (former Intern): *"Mentors here are infinite. You have you're "starting out" mentor for your engineering internship and you're in your group or department and they will definitely feed down the requirements that are expected of you, but as you step out onto that work floor and start getting your hands dirty, you have everyone from recent grads all the way to 30- or 40-year veterans. Overall, the SpaceX internship program is certainly a great training regiment for full time work here or anywhere else in your engineering career."*

Internship Positions

Below is a list of all of the available team internship positions and their descriptions at SpaceX:

Avionics: Designs, develops, manufactures, tests, analyzes and integrates all electronic hardware. This includes but is not limited to: power conversion units, analog and digital circuit boards, FPGA's, communication system units and firmware as well as enclosure and wire harness development.

Build Reliability: Ensures quality production of assemblies and piece parts that flow into the production of the vehicle and proactively improve the vehicle reliability. This group develops standard tools, processes, methods, design adjustments and requirements to ensure production reliability.

Customer Operations And Integration: Serves as the main interface between technical teams and customers to ensure our missions are successful and timely. This group encompasses the Space Operations team, which focuses on revolutionizing spaceflight operations for cargo and crew Dragon missions, as well as future company missions. There are roles in product management, mission management, mission engineering and proposal development.

Flight Reliability: Reviews and certifies the vehicle's test and flight readiness. This team works to maintain safe, reliable practices during operations and to bridge operational and design challenges between test and launch. Flight Reliability interacts with the engineering, production, test and launch teams from the planning stage all the way to launch for the purpose of reviewing the vehicle configuration and status.

Launch Engineering: Fly, recover, fly again. The Launch team works at our Florida launch sites, as well as Hawthorne headquarters on our Falcon 9, Falcon Heavy and Starship launch vehicles. Teams within this organization include but are not limited to: Ground Operations, Payload & Vehicle Integration, Recovery Engineering, Fleet Management, Launch Site Systems and Development Test.

Materials Engineering: Creates materials solutions to expand the capabilities of current vehicles and to push design boundaries for our future ones. They cover a wide range of materials disciplines from alloy development, large scale composite structures, additive manufacturing and traditional aerospace materials. This group's work supports new designs, enabling re-usability of rockets and materials research focused on interplanetary travel.

Production: Leveraging best-practices from a wide variety of industries, SpaceX aims to scale its production rate beyond that of any other launch company in the world. The Production group provides hands-on experience in a fast-paced environment with cutting edge technology. This group includes but is not limited to: Launch Vehicle and Spacecraft Assembly, Tooling Engineering, Industrial Engineering, Automation & Controls and Manufacturing Engineering.

Propulsion: Creates advanced rocket and spacecraft propulsion systems. This team designs, analyzes, builds and tests engines that will one day take humans to Mars. SpaceX currently builds more rocket engines than any other organization in the country. Furthermore, the design and performance parameters of SpaceX engines are notably recognized throughout the industry.

Purchasing: The SpaceX purchasing team is responsible for ensuring lowest product cost and risk through supplier management, strategic negotiations; spend analysis and continuous process improvement. An intern joining this team would expect to gain first-hand experience analyzing supplier and spend data, communicating with suppliers, purchasing direct and/or indirect goods or services, and support of various projects like creating and implementing new processes.

Software: Develops software used to design, develop, test and launch rockets, spacecraft and satellites. In addition to embedded software engineering, they also do simulations, distributed data management and develop analysis tools used in preparation for a launch. Their problem domains span embedded, fault tolerant, flight control, web, mobile, cloud and big data computing. The products that they develop run on low-power space computing platforms, mobile devices, desktop systems and in data centers.

Supply Chain: Improves and develops the business processes that enable SpaceX to design, build, launch and reuse the most advanced rockets in the world. Interns in this group will help innovate and improve efficiencies to ensure that the Supply Chain team is continuously evolving to meet its internal customer demands and to achieve SpaceX's overall mission of sending humans to Mars.

Test Operations: SpaceX tests every engine, component and stage for development and flight at a 4,000-acre state-of-the-art rocket development facility in McGregor, Texas. Interns will receive the opportunity to take ownership of projects related to the systems and hardware used to prove out the next generation of American rockets.

Vehicle Engineering: Designs, develops and qualifies hardware on all vehicles. This group partners with engineering and production teams to create innovative, mass-efficient vehicle and spacecraft structures. This group includes but is not limited to: propulsion components, structures engineering, in-space propulsion, integration & test, vehicle analysis, thermal & fluid analysis, life support systems and guidance navigation & control.

What Qualifications Does SpaceX Look for in an Internship Candidate?

Basic Qualifications:

- Must be currently enrolled at an accredited four-year university or college
- Business operations or software applicants must be currently enrolled

at an accredited university or college or within 6 month post-graduation

Preferred Qualifications:

- GPA of 3.5 or higher
- Strong interpersonal skills and ability to work effectively in a team environment, accomplishing tasks with limited resources at a rapid pace
- Intermediate skill level using Windows Operations Systems
- Intermediate skill level using Microsoft Office (Word, Excel, PowerPoint, Outlook)
- Technical roles: Hands-on experience through engineering project teams, lab research, or through a prior relevant internship or work experience
- Business operations roles: Prior relevant internship or work experience

What are the basic and preferred qualifications to apply for the Associate Engineer Program?

Basic Qualifications:

- Completed a bachelor's degree from an accredited four-year university within the past 6 months by start of employment or currently enrolled in a graduate program

Preferred Skills and Experience:

- GPA of 3.5 or higher
- Strong interpersonal skills and ability to work effectively in a team environment, accomplishing tasks with limited resources at a rapid pace
- Intermediate skill level using Windows Operations Systems
- Intermediate skill level using Microsoft Office (Word, Excel, PowerPoint, Outlook)
- Hands-on experience through lab research, engineering project teams, or through a prior relevant internship or work experience

If your resume and credentials yield you an interview, below are examples of questions SpaceX interviewers have asked prospective SpaceX interns to answer.

The hiring process typically consists of two in-depth phone interviews, during which intern candidates have to answer questions like the following (Walters, 2015):

"What are composites?"

"What is the size of an integer on a 32-bit system?"

"Let's say you have a variable 'var' assigned to be '2'. What will display if you print 'var++'? If you print '++var' on the next line, what will be displayed? What is the final value of 'var'?"

"What is a null pointer?"

"If you have a large, heavy object moving very, very fast, how do you safely slow it down?"

Conclusion

Prior to COVID-19, the Class of 2020 could anticipate a strong job market upon graduation. This has dramatically changed as we head into a recession and high unemployment rates. This tragic experience will, for better or for worse, alter several aspects of our lives going forward.

It's important that college students strongly consider degree programs leading to job opportunities marked as "high need" and acquire skills that will be essential for transitioning the economy from its current state to one that is driven by digital interaction. For example, the requirement to self-isolate has forced innovation upon both industries and educational institutions that have been slow to adopt. Those who have embraced technological advancements will fare much better than their non-technical counterparts. We are already beginning to see this "survival of the fittest" fiercely play out before our eyes, as business that are adapting to the new paradigm are demonstrating resilience, while those that remain reliant on traditional business processes are in rapid decline.

Thus, not *all* college students will experience the negative effects of the recession upon graduation. There will be more job opportunities (and less competition) for college students in STEM programs and especially those who will seek a space-related occupation--which are, and will continue to be, in demand.

Albeit SpaceX is certainly not the only the only company in the space industry offering the added benefit of internships to college students, in terms of future job security, it's apparent SpaceX is poised for growth. And where there's growth, therein lies opportunity.

For more information on the SpaceX Internship https://www.spacex.com/internships

References

Angeles, D. and Vilorio, D. (November 2016). "Space careers: A universe of options," *Career Outlook,* U.S. Bureau of Labor Statistics. Retrieved from https://www.bls.gov/careeroutlook/2016/article/careers-in-space.htm

Cooke, K. (April 20, 2020). "SpaceX Satellite Internet: What you Need to Know about Starlink." *SatelliteInternet.* Retrieved from https://www.satelliteinternet.com/providers/starlink/

Mack, E. (2020, April 1). "Coronavirus pushes SpaceX competitor OneWeb into bankruptcy". *CNET*. Retrieved from https://www.cnet.com/news/coronavirus-pushes-spacex-competitor-oneweb-into-bankruptcy/

Koren, M. (April 13, 2020). "The Mission NASA Doesn't Want to Postpone." *The Atlantic.* Retrieved from https://www.theatlantic.com/science/archive/2020/04/spacex-nasa-coronavirus-astronauts/609871/

Mazumdar, T. (March, 26, 2020). "Coronavirus Update: SpaceX Is Creating Hand Sanitizer, Face Shields To Donate Amid COVID-19 Pandemic." *IBTimes.* Retrieved from https://www.ibtimes.com/coronavirus-update-spacex-creating-hand-sanitizer-face-shields-donate-amid-covid-19-2947073_

Morgan Stanley. (July 2, 2019). "Space: Investing in the Final Frontier." Retrieved from_https://www.morganstanley.com/ideas/investing-in-space

NSR. (June, 2019). Global Satellite Capacity Supply and Demand, 16th Edition. Northern Sky Research. Retrieved from https://www.nsr.com/research/global-satellite-capacity-supply-demand-16th-edition/

Palerm, L. (August 28, 2019). "WILL LEOS CREATE A TRILLION-DOLLAR INDUSTRY?" Northern Sky Research, Retrieved from https://www.nsr.com/will-leos-create-a-trillion-dollar-industry/

Walters, N. (October, 14, 2015). "11 tough interview questions you may have to answer if you want to intern at SpaceX". *Business Insider*. Retrieved from https://www.businessinsider.com/11-tough-spacex-interview-questions-interns-get-2015-10#describe-the-design-process-of-a-series-of-pipes-to-be-used-in-a-rapid-fueling-system-for-a-liquid-propellant-rocket-engine-be-sure-to-include-which-equations-would-be-best-for-the-case-at-hand-for-fluid-pressure-calculations-and-structural-considerations-11

Book Review

Understanding Space Strategy: The Art of War in Space

Title: ***Understanding Space Strategy: The Art of War in Space***
Author: John J. Klein.
Publisher: Routledge, 2019,
Print ISBN: 978-1-1383-54623,
Pages: 245
Price: USD 155.00
Reviewer: Dr. Mark T. Peters II, USAF, Retired

The US Space Force, founded December 20, 2019, recognizes high frontier conflict and poses sufficiently different challenges from land, air, or naval war to require space domain specialists. Winning conflict requires developing effective strategies early and a change from terrestrial to celestial will require extremely specialized strategies. Strategy development traditionally begins with historical founders like Clausewitz, Sun Tzu, and Mahan. Those three, among others, are extensively referenced by John J. Klein throughout his book, *Understanding Space Strategy: The Art of War in Space.* Beginning with strategy basics and unique space elements before presenting four case studies, Klein firmly ties historical approaches to modern conflict. Those with limited strategic backgrounds will find this book immensely helpful. *Understanding Space Strategy* contributes an effective primer, perhaps suitable for newly minted Space Force personnel to link celestial domain possibilities to terrestrial concerns.

The text begins by affirming that conducting a space war does not inherently change a conflict's nature. Strategy, "the art and science of marshaling and directing resources to achieve an objective," follows similar rules regardless of domain (6). Categorizing space-based activities into civil, commercial, intelligence and military areas, each operates through corresponding legal restrictions, behavioral norms, and environmental truths. Nine chapters detail Klein's thoughts with roughly two idea-based sections. The first section, in four chapters, examines strategic, technical, and legal constraints, while the second, in five chapters, applies several strategies to several case studies. Described constraints begin with known limitations, such as the law of war and scarce resources, while finishing with a deterrence model review. Each case examines several states before proposing potential non-military, offensive, or defensive strategy options. Extensively rooted in traditional thinkers who were once controversial, barely a page passes without quoting well-tested, strategic experts.

Klein's first three chapters incorporate strategic concepts, technological impacts, and several deterrence models. An original term, Celestial Lines of Communication (CLOC) appears here after being coined by Dr. Klein in earlier publications. Traditional lines of communication describe material and non-material movement between any two points. The basic discussions suggest ensuring freedom of space, like that of land or air, and subsequent strategic success depends on CLOC control. Primarily terrestrial concepts, including World War II's maritime warfare or nuclear weapons, appear as the historical context to explore similar celestial applications. No unique technological aspects or strategies appear, as generically, technology becomes a supporting conceit rather than a revolutionary one. The final chapter on the law of war features traditional applications, stipulating that orbital debris creation, from kinetic or non-kinetic activity, would be legal during either self-defense or lawful targeting. These initial concepts are the foundation for Klein's subsequent case studies.

Each study includes a two-page historical space summary of the referenced state. Great powers—US, Russia, and China—have proven they can gain and maintain space control based on independent human spaceflight, unrestricted launch, and current orbital operations. Medium powers—India and Iran—can manage independent satellite development and launch. Emerging powers—Canada, Saudi Arabia, and non-state actors—can develop, maintain, and control some functions without organic launch capabilities. The last case integrates independent commercial development, showing that states may require commercial support to maximize capabilities. Each history describes how a state attained previous goals. Beyond broad categorizations, no comparison appears for actual or theoretical conflict between space powers. The great powers descriptions concentrate largely on strategic approaches, while the other two sections are described by past launch activities and orbital control successes. Once establishing each power's preferred celestial operations, the cases discuss generic strategies that Klein feels are possible for the broader power category of emerging, medium, or great.

Each power level study categorizes potential space control strategies as non-military, offensive, and defensive. The commercial section abandons this format to address primarily US opportunities, with a few Chinese references. The strategies are tailored to capabilities but remain defined only as an aggregate rather than addressing specific or past strategic employment. Non-military strategies applied diplomatic, economic, or information/cyber functions toward executing space control through non-orbital functions. Offensive capability ranges from great powers pursuing blockades to medium powers pursuing commerce raiding with no emerging power offensive strategies. Great power defensive strategies include political escalation control, with medium powers favoring guardian systems, and emerging powers seeking protracted war. Terrorist activities against terrestrial control sites have also appeared as an emerging power capability. No strategy appears nationally tailored, as the most detailed reference describes China's "space blockade operations" (kongjian fengsuo zuozhan) for a single paragraph (109). Commercial activities are neither offensive nor defensive but discuss a capability range including data analysis, global imaging, and investment management, among others.

After summarizing these cases, Klein looks toward future space strategies. His pessimistic vision concludes that space war will be inevitable. Settling on inevitability, rather than discussing what future war may consider or how space conflict might support terrestrial engagement, the look instead turns to considerations of space traffic management, mitigating orbital debris issues, and arms control. Many first-world military functions, such as global positioning, intelligence collection, and communication, rely heavily on space, and the potential impacts to those areas do not appear. The non-military topics fail to address how the listed items improve space control and leave the reader concluding that space will be cooperatively managed rather than ruined if one state attempts sole dominion.

The overall review proves effective without offering any new space strategy interpretations. The book serves best as primer, introducing new readers to historical strategic aspects without tailoring the discussion to a single state's integrated strategy or how various strategies conflict during celestial applications. The text offers many, many quotes—about one, one-block paragraph quote for every two pages, from various respected old and new strategists—but fails to deliver new interpretation or integration. Some elements were too historical in seeking strategic ideas about how orbital constraints affect campaign success. For example, the suggestion that medium powers pursue commerce raiding techniques emerged from historical naval analysts Charles Callwell (printed in 1905) and Julian Corbett (printed in 1911). Accurately summarizing century-old naval commerce raiding strategies, Klein fails to describe where the strategy could counter current space control beyond "Affecting an adversary's space reliant business, commerce, and trade can achieve strategic effect" (140). The application here, and throughout, often left me wondering why, where, and how those strategies apply celestially.

Overall, *Understanding Space Strategy* introduces a great number of valuable sources to the beginning strategist, is thoroughly referenced, and is quite comprehensive at the basic level. The exhaustive quotes, chapter endnotes, and detailed bibliographies make the volume a relatively quick read despite the two hundred plus page count. The book should make a good primer for those who have not yet done much reading in strategy or space. Perhaps the new US Space Force could even integrate this volume into their junior officer or NCO training. Lacking any new interpretation for classical strategists and without introducing any unique approaches other than the interesting term CLOC, *Understanding Space Strategy* likely remains best suited to those new to space and strategy fields.

Dr. Mark T. Peters II
USAF, Retired

Book Review

Apollo's Legacy: Perspectives on the Moon Landings

Title: ***Apollo's Legacy: Perspectives on the Moon Landings***
Authors: Roger D. Launius
Publisher: Smithsonian Books (May 14, 2019)
Language: English
Pages: 264
ISBN-10: 1588346498
ISBN-13: 978-1588346490
Price: Hard cover 1858 or Kindle USD 17.99
Reviewer: Arnauld Nicogossian, MD, FACP. Distinguished Senior Fellow, Schar School of Policy and Government, George Mason University.

On December 19, 1972 in the South Pacific, I watched from the deck of USS CVS 14, Ticonderoga[1] three large red, white, and blue parachutes deploy and slowly bring down the Apollo 17 command module to a perfect splashdown off the bow of the aircraft carrier. The NASA team and sailors were exuberant and proud of the country's achievements and were looking for a brighter future for space exploration. Little did we suspect that this was the last lunar crewed mission and that the next forty-eight years would be destined for low Earth orbital missions. The United States (US) was embroiled in a protracted and costly "police action" in Southeast Asia, significant social events were challenging established governance, and the novelty of the Apollo program was relegated to history. Further resources for the Lunar and Mars crewed missions were redirected instead to the US first Orbital Space Station, the US-Soviet Apollo-Soyuz Test Project détente mission, and the follow-up Space Shuttle. The first decade of post-World War II East-West competition for world supremacy created serious concern and a sense of insecurity for US defense and technological supremacy. Regional police actions by the US to counter communism's expansion, the Bay of Pigs disaster, Vietnam entanglement, the loss of the monopoly of nuclear threat, and Soviet achievements in space exploration contributed to the US society's loss of a feeling of security. Was it the loss of interest in the Moon or purely a tribute to the completion of a technical goal establishing the US as a can-do nation of pioneers? In 1962 at Rice University in Houston, Texas, John F. Kennedy charged NASA and the aerospace industry to "... achiev[e] the goal, before this decade is out, of land-

1 USS Ticonderoga, Essex-class aircraft carriers, was built during World War II and was decommissioned in 1973. She participated in the recovery of Apollo 16 and 17 and Skylab 2 missions.

ing a man on the Moon and returning him safely to Earth." NASA delivered on this promise and fulfilled Kennedy's vision, creating in the process many novel commercial opportunities.

Since then, NASA and its proponents have continued to lobby for a return to the Moon, to stay there, and to undertake a journey to Mars. While several US presidents have supported these goals, resources have failed to materialize until recent times.

It is unquestionable that the Apollo program left an everlasting historical and societal impact on the US and the world.

The *Apollo's Legacy: Perspectives on the Moon Landings* was written by Dr. Roger D. Launius in anticipation of the fiftieth anniversary of the first astronaut landing on the Moon. Dr. Launius is an authority on the space program, world-renowned historian, former Associate Director of Collections and Curatorial Affairs at the Smithsonian National Air and Space Museum, and NASA's Chief Historian. Dr. Launius is a prolific writer and frequent speaker at major space events and has appeared on televised space documentary shows. His intimate knowledge of the evolution of the space program makes him a well-qualified expert to critically evaluate the Apollo program's influence on society, policy, politics, and sociocultural evolution.

In this easy-to-read book, Dr. Launius explores many of the events that characterize the Apollo program and its impact on future developments in space and Earth sciences and the continued analysis of the samples returned from the Moon, unravels the mysteries of the origins of the creation of the Solar System and describes the many reasons we are still waiting for future interplanetary exploration. Through the chapters, it is fascinating to follow the evolution of the program, the mourning of the death of the Apollo 1 astronauts, and the world's prayers for the safe return of the Apollo 13 astronauts. The technological impacts of the program have sustained new endeavors, and many of the conversational phrases used by astronauts and mission control have entered our daily lexicon. Dr. Launius also discusses the philosophy and discourses of individuals who deny that the Moon landing even occurred. Many of the events presented at the beginning of the book review are discussed in detail in different book chapters. Dr. Launius also discusses in depth the lessons learned from Apollo and their implications for the future exploration of space.

This book is a valuable resource for students, experts, and teachers of the space sciences and engineering. It will provide readers of all ages an invaluable understanding of the exciting human exploration of space at a time of significant societal and cultural evolution during the Cold War era.

Arnauld Nicogossian, MD, FACP

Book Review

Essentials of Public Health Biology: Biologic Mechanisms of Disease and Global Perspectives

Title: ***Essentials of Public Health Biology: Biologic Mechanisms of Disease and Global Perspectives***

Authors: Loretta DiPietro, Julie Deloia, and Victor Barbiero

Publisher: Jones & Bartlett Learning

Print ISBN: 9781284077919, 1284077918

e-text ISBN: 9781284167825, 1284167828

Edition: First

Pages: 259

Copyright year: 2019

Reviewer: Arnauld Nicogossian, MD, FACP. Distinguished Senior Fellow, Schar School of Policy and Government, George Mason University.

Price: paperback USD 26.69 to 88.95: eBook USD 69.93.

Advances in human genome and availability of large-scale population health databases promise a more rational approach to workers' health, disease prevention, and treatment, based on individual biological variability (Khouri 1996). The environment has become a significant health factor in triggering a genomic response and occasionally inducing pathological changes (Rappaport 2016).

Many vulnerable population subgroups are now afforded screening for diseases such as Tay-Sachs (Benson & Therrell 2010), sickle cell, phenylketonuria, and Thalassemia. Healthcare providers and public health planners are now equipped with clinical assays to screen and identify pre-symptomatic conditions, stemming from gene mutations leading to post-birth, late life potential risks, and forensic testing in legal cases. Nutrition and some medications can alter the intestinal flora, leading to disease. The developing field of nutrigenomics (Comerford & Pasin 2017; Neeha & Kinth 2013) has the potential to promote nutrition to reduce the risk of diet-related diseases. The United States adopted the *Newborn Screening Saves Lives Act* of 2007 (NBSSLA), providing federal guidance to states for newborn screening, parental and provider education, and community outreach, which are examples of such prevention practices.

Since 1963, the field of public health biology has been maturing and many universities are now offering academic training, which brings together the traditional disciplines of epidemiology, population health, prevention, environmental health, and genetics.

Essentials of Public Health Biology is the first edition of a concise, informative, and extensively illustrated textbook, which is supported by web-based teaching materials available to instructors and students. The book is divided into two sections and nineteen chapters.

Chapter formats are standardized, and include key terms, discussion questions and, in many instances, implications to public health. It is well referenced, with up-to-date citations. The authors and contributors are well-published academics in the disciplines of public health and biology.

This much needed textbook is designed as an introductory text for undergraduates and as part of the Master's of Public Health core studies. The authors of the textbook intend to highlight the genetic principles in health and disease and their interactions with environmental factors. This textbook fulfills the authors' aim to bring together the genetic and environmental causation of health risks and promote evidence-based interventions.

The first section of the textbook provides an introduction to the basic genetic and molecular concepts for health and disease. The reader is introduced to the important subject of risk communication and risk management. The second section deals with communicable and non-communicable diseases of major concern to public health practitioners. Future editions of this textbook should include a stand-alone chapter on ethical considerations in the use of genetics in public health, especially addressing topics such as discrimination based on the epidemiological information, and the protection of large-scale health information databases; these topics would markedly enhance the textbook.

Arnauld Nicogossian, MD, FACP

References

Benson Jane M., & Bradford L. Therrell Jr. 2010. "History and Current Status of Newborn Screening for Hemoglobinopathies." *Semininars in Perinatology* 34 (2): 134-44.

Comerford, Kevin B., & Gonca Pasin. 2017. "Gene-Dairy Food Interactions and Health Outcomes: A Review of Nutrigenetic Studies." *Nutrients* 9 (7): 710.

Jeanmonod, Rebecca, & Donald Jeanmonod. 2019. "Inborn Errors of Metabolism." In *StatPearls*. Treasure Island: StatPearls Publishing. https://www.ncbi.nlm.nih.gov/books/NBK459183/.

Khouri, Muin J. 1996. "From Genes to Public Health: The Applications of Genetic Technology in Disease Prevention." *American Journal of* Public Health 86 (12): 1717–22.

Neeha, V. S., & Priyamvadah Kinth. 2013. "Nutrigenomics Research: A Review." *Journal of Food Science and Technology* 50 (3): 415–28.

Prüss-Ustün, Annette, Emilie van Deventer, Pierpaolo Mudu et al. 2019. "Environmental Risks and Non-Communicable Diseases," *BMJ* 364: l265.

Rappaport, M. Stephen. 2016. "Genetic Factors are not the Major Causes of Chronic Diseases." *PLoS ONE* 11 (4): e0154387. https://doi.org/10.1371/journal.pone.0154387.

This publication is available open access at:
http://www.ipsonet.org/publications/open-access

Thanks to the generosity of the American Public University System

APU American Public University

AMU American Military University

Made in United States
Troutdale, OR
11/03/2026

56845242R00126